Filtering Media by Electrospinning

Maria Letizia Focarete • Chiara Gualandi
Seeram Ramakrishna

Editors

Filtering Media by Electrospinning

Next Generation Membranes for Separation
Applications

 Springer

Editors
Maria Letizia Focarete
Department of Chemistry "Ciamician"
University of Bologna
Bologna, Italy

Chiara Gualandi
Department of Chemistry "Ciamician"
University of Bologna
Bologna, Italy

Seeram Ramakrishna
National University of Singapore
Singapore, Singapore

ISBN 978-3-319-78162-4   ISBN 978-3-319-78163-1   (eBook)
https://doi.org/10.1007/978-3-319-78163-1

Library of Congress Control Number: 2018942219

Printed on acid-free paper

This Springer imprint is published by the registered company Springer International Publishing AG part of Springer Nature.
The registered company address is: Gewerbestrasse 11, 6330 Cham, Switzerland

Contents

Chapter 1
Electrospinning Technology for Filtering Membranes Fabrication

Vincenzo Guarino and Alessio Varesano

Abstract Filtration is simple, versatile, and economical means to recruit and confine sub-micrometric or nanometric particles and ions, that is, aerosol, water, and air pollutants—in order to provide a highly efficient process for air/water cleaning. Indeed, the growing amount of toxic particles with a large specific surface area, mainly related to increasing combustion emissions in the external environment, is drastically reducing population life quality, due to the ease of crossing human body barriers—through inhalation and absorption. In the last two decades, rapid improvements in nanotechnology are opening new perspectives for the development of innovative filters with permeability properties able to more efficaciously prevent the release of nanoparticles in the atmosphere, soil, or water environments. In particular, electrospinning is emerging as one of the most versatile and cost-effective technologies to design fibrous filtering devices, capable of removing sub-micrometric/nanometric-sized pollutants from gas streams, drastically improving the current efficiency of commercial filters (up to 90–99%). Herein, an overview of the current technological strategies based on electrospinning will be introduced in order to discuss future challenges and trends in the use of electrospun membranes as innovative nano-porous membranes for filtering applications.

1.1 Introduction

To date, a large variety of heavy metals have been used in the manufacturing processes of various industries. The exceeding and uncontrolled release of ions in effluents or volatile nanoparticles into the atmosphere currently represents a main cause of severe damage to human health [1]. Air pollution due to particulate matters and gaseous pollutants may cause asthma, nausea, skin irritation, high blood pressure,

V. Guarino (✉)
Institute of Polymers, Composites and Biomaterials, National Research Council of Italy,
Naples, Italy
e-mail: vguarino@unina.it; vincenzo.guarino@cnr.it

A. Varesano
Institute for Macromolecular Studies, National Research Council of Italy, Naples, Italy

© Springer International Publishing AG, part of Springer Nature 2018
M. L. Focarete et al. (eds.), *Filtering Media by Electrospinning*,
https://doi.org/10.1007/978-3-319-78163-1_1

cancer, birth defects [2], respiratory and cardiovascular diseases [3] with health hazards depending upon exposure level and pollutant nature [4]. Alternatively, heavy metal ions can easily be mixed into the water reservoir—acting as a carrier—which distributes metal ions to the surrounding environments but effective routes to separate metal ions from the reservoir water are still partially unexplored [5].

Fibrous filters may represent simple and economical devices able to efficiently remove particles at micro/nanometric-size scale from different fluid (i.e., air, water) streams. More in general, particles could be retained by the fibrous fiber network when they collide and attach onto the fiber surface by mechanical (i.e., inertial impaction, interception, Brownian diffusion, gravitational settling) or electrostatic deposition mechanisms [6]. As a consequence, filters have to present an intricate mat of fine fibers arranged in a way that most are perpendicular to the direction of airflow, with porosity from 70% greater than 99%, and fiber range in size from tens of nanometers up to 100 micrometers. Usually, small fiber diameter increases filtration efficiency but this is coupled with high pressure drop irrespective of the particle size, so negatively influencing quality factor. Indeed, nanofibers with 185 nm as diameter showed higher quality factor than fibers with 94 nm as diameter [7], due to the ability of relatively larger fiber meshes to promote the diffusion of 50–90 nm particles. In the case of bigger particles (100–380 nm), interception mechanisms dominate over diffusion, so that small fiber diameters reduce the effective pore openings, thus enhancing the direct-interception effect for particle capture [8]. Hence, depending upon the target particle size, they have to be variously engineered with characteristic mesh sizes— that is, coarse (2.5–10 mm), fine (0.1–2.5 mm), and ultrafine (<0.1 mm) [9, 10], as a function of the specific application area (i.e., disposable respirators, industrial gas cleaning equipment, cleanroom air purification systems, automotive cabin air filters, and indoor air purifiers) [11].

With the increasing knowledge in the field of nanotechnology, many techniques are emerging for the synthesis and processing of fibrous materials at the sub-micrometric/nanometric level. Among them, electrospinning is considered to be one of the most efficient techniques used for the fabrication of micro/nanostructured platforms for various potential applications in biomedical area, water/gas filtration, desalination, and sensors/biosensors. Main advantages of electrospun fibrous membranes for filtration are mainly ascribable to their small pore size and high specific surface area. The average pore size can be 4–100 times smaller than that of micro-fiber membranes [12] which can capture dust particles on their surface, ultimately improving filtration efficiency [8]. The specific surface area can be several times higher than that of conventional microfibers, due to the presence of micropores—sizes around 2 μm—and mesopores—ranging from 2 to 50 nm—in the fiber network [13], thus allowing to significantly improve filtration efficiency, only by using thin layers of electrospun fibers [14]. More interestingly, electrospun nanofibers may be used to incorporate surface-active chemical compounds to improve fiber functionality at the nanoscale without deteriorating the breathability or moisture vapor diffusion, with relevant advantages in terms of comfort of personal protective clothing [15]. The possibility to process a wide variety of layered structures made of different materials, including glasses, plastics, ceramics, metals, and also biomaterials such as polysaccharides (i.e., cellulose derivatives, chitosan [16]), proteins (i.e., wool keratin [17]), and biodegradable polyesters (i.e., polylactic acid (PLA) [18]), allows fabricating smart filters able to solve health problems associated with bulky respiratory filters such as sweating, thermal stress, and breathing difficulty [19].

Recent studies have reported that electrospun fabrics may be not only assembled in the form of pristine polymer nanofibers but also developed starting from synthetic and natural polymers (i.e., keratin/PA [16, 17]) to obtain blends or polymer composites with uniform nanometric diameters. They may be successfully used in active filtration, for absorbing/removing toxic substances dispersed in air (i.e., formaldehyde [20]) or retaining heavy metal ions from contaminated water, thanks to their unique morphological properties (i.e., surface-to-length ratio, interconnected porosity, surface porosity), which drastically improves the performance with respect to conventional materials.

Herein, it is proposed an overview of current use of electrospinning techniques for the fabrication of filtering membranes, by describing, firstly, basic working principles and, secondly, different technological strategies and setup configurations currently optimized to design fibrous fabrics for different filtering applications. Lastly, various research challenges and future trends of electrospun membranes in fluid and particle filtration will be discussed.

1.2   Electrospun Fibers for Filter Fabrication

1.2.1   Electrospinning by Polymer Solution

Solution electrospinning is based on the application of a high potential electric field to a polymer solution to form small-diameter fibers at micro or sub-micrometric-size scale. Their peculiar morphological features have found different use in diverse scientific fields, including electronics, for anode components in lithium ion batteries [21], micro/nano-devices [22], and high-performance air filters with high filtration efficiency and low air resistance [23, 24].

The experimental setup is generally simple. Using a syringe pump, the polymeric solution is forced through a metal needle to form a pendant drop at the tip of the capillary. High voltage potential is applied to the polymer solution, thereby inducing free charges into the polymer solution, able to transfer tensile forces to the polymer liquid, charged in response to the applied electric field [25]. When the applied potential reaches a critical value required to overcome the surface tension of the liquid, polymer drop takes a cone-like projection, known as Taylor's cone, and a liquid jet is ejected from the cone tip [26]. Electric forces then accelerate and stretch the polymer jet, causing the diameter to decrease as its length increases. After the initiation from the cone, the jet undergoes a chaotic motion or bending instability and as it goes through the atmosphere toward the opposite charged collector, the solvent evaporates, leaving behind a dry fiber on the collecting device

[27]. Meanwhile, additive instability phenomena may be induced by the coupling of the liquid strand with the electric field (e.g., axisymmetric instability) providing a statistical variance of the jet radius that causes a modulation of the surface charge density. This variation, in turn, generates tangential forces, which couple to the radius modulation and amplify it, thus forming quasi-spherical defects or beads along the fiber-like pearls on a string [28]. Apart from the surface tension and forces of the polymer solution that is electrospun, there are various other processing variables that are key factors to reach the final performance of the nanofibers. In the last 20 years, many researchers have investigated the effects of various processing variables on the structural morphology and properties of electrospun fibers, in order to find relative correlations between various processing parameters and their impact on the fiber morphology and surface and bulk properties. First studies mainly refer to the identification of a critical voltage value to trigger the polymer fiber ejection. This critical value varies with the type of polymer solution as a function of polymer molecular weight, solvent permittivity, and solution concentration [29, 30]. When electric field is weaker or stronger than this critical value, beads and defects may occur along fibers as a consequence of the presence of instability phenomena. Current studies shows that the increase in voltage leads to a stretching in fiber length and a decrease in fiber size [31]. In particular, as the voltage increases, multijet phenomena may occur, thus generating fibers with different diameters [32]. Another effective parameter on physical and chemical properties of produced fibers is polymer flow rate. Change in the polymer flow rate causes change in fiber morphology, with an increase in fiber diameters and beads occurrence as the flow rate increases [33]. Morphology, structure, physical, and chemical properties of electrospun fibers may be also affected by the nozzle-to-collector gap. Indeed, it directly influences evaporation rate and deposition time, so particularizing fiber size and surface morphology. As the distance between the nozzle and the collector decreases, electrospun fibers tend to be wet, due to the presence of residual solvent which, ultimately, promotes a reduction of fiber mesh size and fiber flattening [34].

In the case of solution electrospinning, a key aspect certainly concerns the optimization of process environmental conditions generated by the presence of exceeding solvent into the fiber body. This aspect is extremely important to be considered, not only for safety reasons, but also for the quality of the final products strictly related to the solvent residues trapped inside the bulk nanofibers [35]. Whilst solvent traces are not fully invasive on fiber morphology and functionality in the case of non-textile applications (e.g., nano-electronics), instead, solvent contribution may become crucial in the case of biomedical or filtering applications where an accurate control over solvent residuals is strictly required to preserve the biosafety of the final membrane. In order to remove as much as possible the undesired presence of volatile organic solvent compounds, the use of bioprocesses based on the use of green materials, that is, polymer dissolution in aqueous or less invasive solutions, is emerging as an attractive solution for the fabrication of a new generation of filtering membranes [36].

1.2.2 Other Processing Routes

In the last years, different processing routes based on the application of electrostatic forces on polymer materials have been explored to form fibrous membranes with micro- and/or nanoscale structure. Different driving forces, that is, temperature, pressure, may be applied alone or in combination with electrostatic field to influence the breaking mechanisms able to form polymeric fibers (Fig. 1.1). Among them, melt electrospinning is a temperature-dependent process based on the production of fibers at the micrometric-size scale from melt polymers. With respect to a conventional melt-spinning setup, high voltage is applied and melt polymer and fiber is formed by the interaction of electrical forces and polymer chains, over a critical threshold voltage. Similarly to the basic electrospinning from polymer solution, the mechanism is based on the competition between electrostatic forces, surface tension, and viscoelastic properties of the melt flow, and the fiber features are basically controlled by well-known process parameters (i.e., electrospinning voltage, needle tip–to-collector distance, flow rate) [37]. Flow rates are typically lower than those of polymer solution in the case of solution electrospinning, due to the effect of higher viscosity of melt polymers, thus resulting in a strong limitation for fiber drawing. In contrast to solution electrospinning requiring organic solvents to dissolve the polymer into a homogenous solution, forming thin fibers by solvent evaporation [38–40], in melt electrospinning the fiber is formed by a solvent-free method based on the natural cooling of the melt jet ejected from the charged flow over the polymer melting point, under the electric field application [41]. Hence, this process is not suggested for the encapsulation/release of biolabile/biodegradable molecules (i.e., proteins, polysaccharides, nucleic acid) while no specific restrictions may be ascribable in the case of solution electrospinning [42], despite the fact that some problems may be still registered as a consequence of molecular interactions with aggressive solvents. Besides, melt electrospinning is a safer and "greener" approach than solution electrospinning, not presenting residual solvent release, thus resulting in greatly being used for filtering applications. More interestingly, the melt

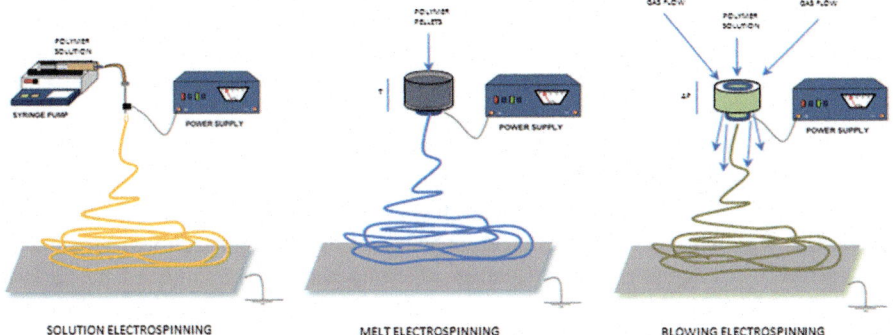

Fig. 1.1 Scheme of electrospinning processing modes

electrospinning process setup may be also improved to work at higher operational rates (ca. 1000 m/min) with respect to conventional ones, that is, by using a new supersonic pump able to work over 3000 m/min (high-speed melt spinning)—thus assuring higher fiber productivity [43].

Alternatively, melt-spun fibers deposition in combination with moving collectors, that is, melt electrospinning writing (MEW), may be successfully used for a layer-by-layer fabrication of membranes with specific features [44] and different kinds of fibers arrangement [45] able to improve the filtration capability of the membrane [46]. This has been proved by Homaeigohar et al., which has studied the main benefits of using spun-bound and melt-blown electrospinning techniques to form polyethersulfone (PES) and polyethylene terephthalate (PET) multilayer electrospun membranes for water filtration. They concluded that PES electrospun nanofiber membranes exhibited a high permeability for pure water to particles with sizes over 1 μm, with a slow decrease of permeability as feed pressure increases [47].

Solution electro-blowing is one of the most feasible techniques to fabricate nanoporous membranes by the use of compressed air combined with electrical field as the driving force to form fibers. The most widely used and simple configuration is composed of a concentric nozzle system, once polymer solution is pumped into the inner nozzle and high-pressure gas is delivered through the outer nozzle simultaneously [48]. In this case, polymer solution is broken into multiple strands under the stretching forces from the surrounding high-speed gas and the combined effect of applied electrostatic forces. In this case, high shear forces at the gas/solution interface concur to the solvent evaporation, controlling fiber diameter by influencing the jet blowing mechanism [49]. More in detail, fiber morphology may be controlled by changing solution injection rate, the gas flow pressure, the nozzle-to-collector gap, and the protrusion length of the inner nozzle [50]. By comparing nanofibers obtained from solution electro-blowing and electrospinning methods, it is evident that similar nano-microfibers may be obtained by similar polymer solutions [49]. However, production efficiency of the blowing process is drastically higher, due to the use of solution feeding rate which is one order of magnitude higher than that those of solution electrospinning one. Another important factor concerns the environmental conditions. Zhuang et al. recently demonstrated that air temperature in the spinning line drastically influences solvent evaporation and, ultimately, cellulose fiber formation via solution blowing. Indeed, by making faster solvent evaporation, temperature increase also reduces the number of thinner fibers, thus nullifying the attenuation of blown streams [51].

As a consequence, solution electro-blowing is a really suitable method for the production of nanofibrous membranes for filtration and separation applications. By a fine control of fiber size and exposed surface, it is possible to improve the exchange properties which are extremely important for gas filtration applications. Most studies currently focus upon the fabrication of composite fibers activated by carbon nanotubes [52, 53]. For instance, polyacrylonitrile (PAN) has been adopted to process composite nanofibers by using specific precursors able to activate the in situ synthesis of carbon nanofibers with 200 nm as diameter and surface area of 2921.2 m^2/g. As a consequence of these peculiar fiber features, CO_2 adsorption results significantly higher than the adsorption of commercial activated carbons—2.70 mmol g^{-1} with respect to 0.734 mmol/g^{-1} adsorbent for commercial products (NORIT R2030), so confirming their potential use for smart textile and filtering use.

1.2.3 Optimization of Spinnerets and Collectors for Large-Scale Production

Within the environment protection area, filtration and purification, materials based on nanofibrous mats are undoubtedly the currently most industrially advanced products, with a worldwide distribution of supplying companies. For instance, textiles in general can benefit from embedding high-tech nanofibrous features for controlling surface properties. In this context, the capability to reach high production volumes still represents the most important requirement for industrial use. The most relevant technological advances that have, to date, been explored to increase the volume of production are mainly based on the modification of the polymer injection system and consist in the introduction of multi-spinneret components that allows parallel multiprocessing in order to extend the final free surface of membranes.

In particular, multi-jet electrospinning based on multi-spinneret components can be arranged by different setups—either in a uniaxial configuration or in a circular geometry—to obtain the fabrication of fibrous mats made of single or multiple materials combined into the same fiber or separately in different ones, with a high control of membrane thickness and improved process throughputs [54]. In order to overcome drawbacks of the multi-spinnerets related to the alteration of the electric field profile induced by the presence of other electrospinning jets nearby, additional auxiliary electrodes with charge of either the same or opposite polarity with respect to the electrospinning jet may be used [55]. Alternatively, the use of secondary electrodes also concurs to improve the fiber production rates in the case of multiple-nozzle spinning systems [56]. In this case, jets number can vary from 2 to 16, by the application of both negative and positive applied bias currents to finely control the basic inter-jet repulsion mechanisms occurring during the fiber collection, in order to provide a more extended deposition area. Recent studies have demonstrated that dispensing systems composed of multiple holes with controlled diameters, drilled along the wall of rotating plastic tubes, may allow producing up to 0.5 g/h of electrospun fibers, with the chance to control the production rate by modifying tube length and spinning holes unit number [57]. More recently, industrial companies have developed a commercially available dispensing apparatus able to produce fibers with a diameter down to 50 nm with a throughput area up to daily thousand m^2 and production rates of several hundred g/h up to some kg/h, mainly suitable for the fabrication of nanofiber coatings.

Indeed, traditional setup configurations based on the use of needles as dispensing systems generally present important drawbacks due to the tendency of polymer solution to form clogs at the spinneret nozzle, with relevant limitations to achieve high throughputs in continuous production processes [58]. As expected, this effect is more evident as the needle diameter decreases, and may be further amplified by the use of high polymer concentrations or multicomponent solutions (i.e., polymer blends, polymer/nanoparticle mixtures). Thus, needleless dispensing systems are currently emerging as a performing element of industrial processing configurations with improved upscaling potentialities. In this case, the electrospinning jet is formed from the free surface of a liquid, without using needles or nozzles, starting from micro-sized droplets, attracted and swiftly deposited in the form of fibers onto the collector target [59]. Recently, the use of rotating collectors with innovative design, that is, helically probed rotating cylinders, may offer several advantages over syringe-based and needleless electrospinning in terms of productivity (six times higher) and processability, not only covering conventional processing strategies but also introducing new benefits for innovative approaches (i.e., colloidal electrospinning) [60].

Despite alternative electrospinning setups being continuously optimizing (i.e., centrifuge electrospinning, rotary electrospinning) to produce nanofibers exhibiting even thinner diameters—less than 80 nm—and increasing throughputs—ca. 1000 times higher than conventional needleless dispensing nozzles—main limitations still concern the control of fiber diameter and the possibility to implement an advanced setup for the fabrication of complex structural organization (i.e., core-shell) and peculiar fiber assembly (i.e., uniaxial alignment, controlled patterning) [61].

In order to properly control fiber assembly, various modifications have been also assessed to the dynamic collector geometry (Fig. 1.2), specifically to reach the uniaxial orientation of nanofibers. The first kind of collectors is referred to the use of two or more parallel electrodes able to promote an excellent alignment along the empty spaces formed between adjacent electrodes by a basic mechanism of electro-

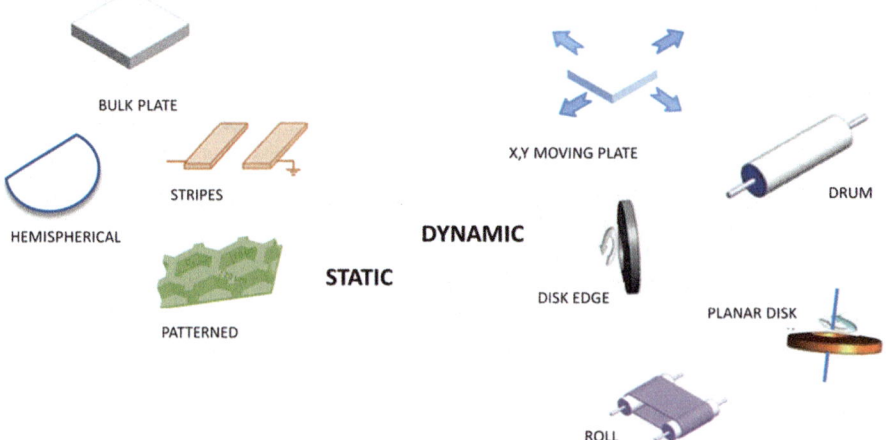

Fig. 1.2 Scheme of static and dynamic classes of collectors for electrospinning process

static rebuttal between fibers [62]. In this most simple case, fiber alignment may be imparted by two metal stripes separated by a well-defined air gap, which promote the stretching of nanofibers across the gap to form a parallel array, under the effect of electrostatic interactions [55]. More recently, several studies also demonstrated that physical properties of electrospun fibers, that is, crystal morphology, molecular orientation, etc., may be affected by the nature of collectors as a function of their specific working mechanism [63]. For this purpose, the most commonly used collector is the rotating drum collector, which allows controlling not only fiber alignment but also fiber diameter as a function of the rotating speed [64]. As a function of drum collecting surface and the inertial forces working during the rotation, the capability of collectors to impart a preferential orientation to nanofibers may change. For instance, rotating disk, that is, a drum with reduced surface available to fiber deposition, is preferentially used to create uniaxial aligned fibers on a nanometric-size scale by confining the collected fibers onto the disk edge [65]. Meanwhile, the collector speed can improve the crystal orientation of fibers due to polymer molecular chains' alignment in the direction of fiber axis which is obtained due to the effect of evaporation mechanisms [66], thus contributing to the ultimate mechanical properties of the fibers [67].

More recently, micro- or macroscale patterns of nanofibers may be obtained also by using conductive grids or a series of charged needles to guide the collection of fibers along the charged regions [68]. For instance, Li et al. developed a variety of aligned crossbar patterns using a layer-by-layer approach with series of conductive grids with void gaps [69]. Zussman et al. attached an aluminum table to a rotating disk collector to generate crossbar patterns in a one-step process [70].

Alternatively, the use of an insulating collector with selective conductive regions may be useful to draw complex patterns by increasing the charge accumulation at the collector [71]. In this context, Ding et al. have proposed the selective deposition of nanofibers on an elastomeric substrate containing gold-coated pyramidal spikes [72]. Star and radial fiber patterns have also been prepared using enamel-patterned steel-sheet collectors [73]. Finally, doped chlorinated polystyrene (PS) and polypropylene with an ionic liquid able to enhance the charge accumulation of the collecting fibers have been used to create a bird's nest pattern of fibers by using an un-patterned plate collector [71].

1.3 Strategies to Design Complex Micro- and Nano-porous Membranes

A typical electrospun nanofiber-based structure offers specific features, such as a large surface-to-volume ratio, a small pore size, and a highly interconnected fiber arrangement, which are crucial characteristics of membranes for many different operations dealing with both physical separation or chemical sequestration processes for gaseous or liquid streams. Electrospun nanofibers are widely proposed for producing membranes for pressure-driven operations, such as micro-, ultra-, and nano-filtration processes and reverse osmosis, temperature gradient operations, such as membrane distillation, and electric potential gradient operations, such as electrodialysis and electrofiltration.

Membrane separation requires complex membrane architectures which can be attained using layers with different densities and porosity (e.g., pressed nanofibers, films, nanoparticles), chemical surface modification (e.g., super-hydrophobic layers), internal nanocomposite structures (e.g., polymer blends, organic-inorganic hybrid structures), for instance.

The membrane shape required by some operations (usually different from the as-spun nanofibrous structure) can be obtained using different strategies, such as unconventional electrospinning techniques (coaxial electrospinning, dual electrospinning) or post-treatments (physical, chemical, or both).

Coaxial electrospinning was used to produce fibrous membranes for efficient oil–water separation [74]. Cellulose acetate and polyamide acid were coaxially electrospun obtaining polyamide acid core/cellulose acetate shell nanofibers. Then the polyamide acid core was imidizated at high temperature to produce polyimide. The cellulose acetate on the surface of the fibers was modified with fluorinated polybenzoxazine and silica nanoparticles. In this case, coaxial electrospinning is useful to produce a membrane with higher tensile strain compared to fibrous membranes composed of pure cellulose acetate, with hydrophobic and high stability in a wide range of pH. In the work, the setup consisted of a needle (0.51 mm of inner diameter) coaxially placed inside an outer needle (1.2 mm of diameter). The solutions were polyamide acid dissolved in N,N-dimethylacetamide for the core feeding, and cellulose acetate dissolved in a mixed solvent of dichloromethane/acetone (3:1 v/v) for the shell feeding. The flow rates were 0.5 and 1 mL/h for the inner and outer needles. The working distance was 13 cm and the voltage was +25 kV.

In another study [75], hydrophobic titanium dioxide nanoparticles functionalized by fluorosilane were incorporated into electrospun membranes using single, coaxial, and dual electrospinning to develop complex membrane configurations for membrane distillation. Electrospinning operating conditions were different depending on the kind of process. In particular, (single) electrospinning was carried out at +18 kV of voltage, 1 mL/h of flow rate, and 15 cm of working distance. In coaxial electrospinning, the voltage was reduced to +14 kV and flow rates were 0.4 and 0.5 mL/h for core and shell (the same solution of core with hydrophobic titanium dioxide nanoparticles), respectively. In dual electrospinning, the voltage applied to the solutions (with and without hydrophobic titanium dioxide) was +18 kV for both, while the flow rates were 0.8 mL/h for the solution with hydrophobic titanium dioxide nanoparticles and 1.0 mL/h for the solution without nanoparticles. The membranes produced with single and coaxial electrospinning showed contact angles close to 150°, whilst those produced with the dual electrospinning were composed of bead-on-string fibers achieving a contact angle of 153.4° and super-hydrophobicity characteristics. Moreover, dual electrospinning produced fibers with different sizes that reduced the membrane porosity because thin fibers fill the space between thick fibers. Another advantage in using dual electrospinning was an effective increase of the amount of exposed particles in the membrane. In turn, the membranes produced by dual electrospinning showed the better membrane distillation performances.

Super-hydrophobic nanocomposite fiber membranes were produced by electrospinning using poly(vinylidene fluoride) and modified silica nanoparticles [76]. The nanoparticles were prepared by a sol–gel method using different silane coupling agents (i.e., n-octyltriethoxysilane, vinyltrimethoxysilane, and vinyltriethoxysilane). Silica particles increased the roughness of the fiber surface compared to the membranes without nanoparticles. In particular, a micro/nano dual-scale structure was formed, which was responsible for the super-hydrophobicity of the nanocomposite fiber membranes. Contact angle is affected by both the coupling agent used and the amount of silica nanoparticles. The highest contact angle (160°) was achieved with n-octyltriethoxysilane coupling agent and an amount of 5 wt% of silica nanoparticles.

Thermally stable hydrophobic nanofibrous membranes were obtained by electrospinning silica/polydimethylsiloxane hybrid fibers [77]. Silica sol and polydimethylsiloxane were merged in dichloromethane producing a solution suitable for a transformation into ultrafine fibers by electrospinning. No change in hydrophobicity was found in flame-treated membranes. The results clearly confirm that the hydrophobic feature is thermally stable. However, the hydrophobicity seems to be related to the density of the membrane: Excellent hydrophobicity was found in dense parts of the membrane, while hydrophobicity may disappear in loose parts.

Graphene was used in combination with poly(vinylidene fluoride-co-hexafluoropropylene) in order to produce super-hydrophobic electrospun nanocomposite membranes [78]. The paper shows that a graphene loading of 5 wt% significantly enhanced the membrane structure and properties. Electrospinning was carried out at +10 kV of voltage, 10 cm of working distance, and 1.0 mL/h of flow rate.

Contact angles of nanofiber membranes can be tuned by mixing different polymers in the electrospinning solution, instead of introducing fillers, such as nanoparticles. An example is the work of Ren et al. [79] where super-hydrophobic polydimethylsiloxane/poly(methyl methacrylate) nanofibers were electrospun successfully producing a membrane with a contact angle of 163°.

Polysulfone/$NiFe_2O_4$ nanostructured composite fiber membranes were produced by electrospinning to be used as a magnetic sorbent for oil removal from water [80]. The $NiFe_2O_4$ content was 30 wt% on the polymer, and the electrospinning conditions were an applied voltage of +25 kV, a working distance of 15 cm, and a flow rate of 0.75 mL/h. The sorption performances of the composite nanofiber membranes were 9.20 g/g for dodecane and 15.11 g/g for motor oil. The sorption capacities were lower than the sorption performances of pure polysulfone fiber membranes electrospun at the same conditions. The decrease in sorption capacity has been explained by taking into account the significant content (30%) of the inorganic material. The advantage of the electrospun composites relies in a facile magnetic separation of the sorbent from the oil/water system.

Multilayered membranes, combining different properties of several polymers, show better performances in specific applications compared to membranes constituted by only one polymer.

An unconventional electrospinning setup was used to produce super-hydrophobic membranes with high water flux to be used in membrane distillation [81]. The setup consisted of three needles in a linear arrangement. The needle in the middle was fed with a solution of poly(acrylonitrile) in N,N-dimethylformamide, and the other two needles with an emulsion of poly(vinyl alcohol) and poly(tetrafluoroethylene) in water. This setup was chosen in order to increase mechanical properties of the resulting membranes because poly(tetrafluoroethylene)/poly(vinyl alcohol) nanofibers had shown poor mechanical strength, and poly(acrylonitrile) nanofibers successfully acted as reinforcement.

A different strategy was used to produce bipolar membranes, which consists of cation-exchange and anion-exchange layers separated by an "inert" intermediate layer. Bipolar membranes are typically used to split water into the protons and hydroxide ions using an electric field. In this work, [82] sulfonated poly (phenylene oxide), polyethylene glycol, and quaternized poly (phenylene oxide) were continuously electrospun on a rotating drum, so that the resulting membranes were composed of several layers of different materials. After electrospinning, the membranes were hot pressed in order to obtain dense structures. The thickness of each layer can be easily tuned by controlling the electrospinning parameters for each solution.

An electrospun multilayer structure with polyamide 6 and poly(trimethylene terephthalate) [83] was also proposed for low-pressure air filtration. The average sizes of the electrospun nanofibers were different for the two polymers. In particular, polyamide 6 nanofibers had an average size of 106 nm, while poly(trimethylene terephthalate) nanofibers had 907 nm. Therefore, the pore size of the polyamide 6-nanofibrous layer resulted smaller than that of poly(trimethylene terephthalate). Benefits were obtained in terms of filtration efficiency using multilayered membranes instead of single-layer membranes.

In order to increase the nanofiber interconnection, in a paper [84], polyethylene terephthalate nanofibers were collected on a rotating drum by electrospinning at the same time two solutions with different polymer concentrations (namely, 20 and 22 wt%) in trifluoroacetic acid/N,N-dimethylformamide (80:20) from two needles (so-called dual electrospinning). The electrospinning parameters were a flow rate of 1.2 mL/h, a voltage of +25 kV, and a working distance of 12 cm for both the needles. The resulting membranes were further post-treated and proposed for an osmosis process.

Another paper [85] reported the use of a patterned collector in order to produce a dense-sparse nanofiber structure for gas filtration. The membranes were obtained by electrospinning polyacrylonitrile nanofibers on an embossed paper and an air bubble wrapping, and compared to a traditional aluminum foil, as collectors. On the aluminum foil, the distribution of the nanofibers is uniform in thickness, while changes in fibre density may be ascribable to the features of patterned collectors. In particular, nanofibers concentrated on the edge of the template that caused different densities of the membranes. Interestingly, the patterned collectors did not alter the nanofiber sizes. Therefore, changes in pore size and porosity were attributed to the patterned collectors only. Efficiency of nanofiber membranes with patterned structures slightly decreased, but the pressure drop significantly decreased, too. As a result, the overall performances of the membranes greatly improved.

Zhang et al. [86] reported the possibility of producing two-dimensional (2D) "nano-nets" with an ultrathin diameter (~20 nm) in a one-step preparation of electrospinning. The nanoscale diameter of these fibers can facilitate the "slip effect" of air molecules, and the resulting small pore size can sieve ultrafine particles from air and gasses. Electrospinning poly(m-phenylene isophthalamide) membranes with bimodal fiber diameters were spontaneously produced. Few randomly distributed nano-wires with a diameter of 30–50 nm connecting regular-sized nanofibers (100–300 nm). This phenomenon was attributed to the high conductivity of the solution that led to a high charging on the flying jet resulting in its branching due to a high electrical repulsion inside the jet itself. For having this result, humidity seems the most important parameter since humidity is linked to the solvent evaporation and charge dissipation.

Nanofibers were collected on a hollow braided polyester rope as a support in order to produce tubular membranes for water and wastewater treatment [87]. Membranes were produced by electrospinning a solution of polyacrylonitrile in N,N-dimethylformamide. Average pore size ranged from 300 to 400 nm. The membrane showed two–three times higher flux than a commercial membrane with a similar pore size. High rejections were obtained for particles with 1–3 mm diameter. However, the commercial membrane exhibited higher efficiency for fine particles (500 nm and 100 nm). It seems that the pore size of the electrospun membrane at the first layer was larger than that of the commercial membrane.

1.4 Fiber Membrane Post-treatments

Post-treatments often involve a high temperature step after electrospinning (curing, sintering, hot pressing, heat bonding) in order to change the chemical structure and/ or morphology of the nanofibers. Other post-treatments are related to multiple material depositions (Fig. 1.3). In this case, nanofibers are used as a support for other layers that can include nanofiber multilayers, film deposition (for instance, by interfacial polymerization), or nanoparticles deposition (for instance, by electrospray). Chemical modifications of the nanofiber surface are often produced by grafting, but also thermal treatments, such as sintering and curing, can change chemical groups of the nanofiber surfaces.

1.4.1 Physical Modifications

Combinations of electrospinning and electrospray were used to change the surface morphology of nanofibers in order to enhance the hydrophobic behavior of the membranes for a variety of applications. Ryu et al. [88] produced layers against

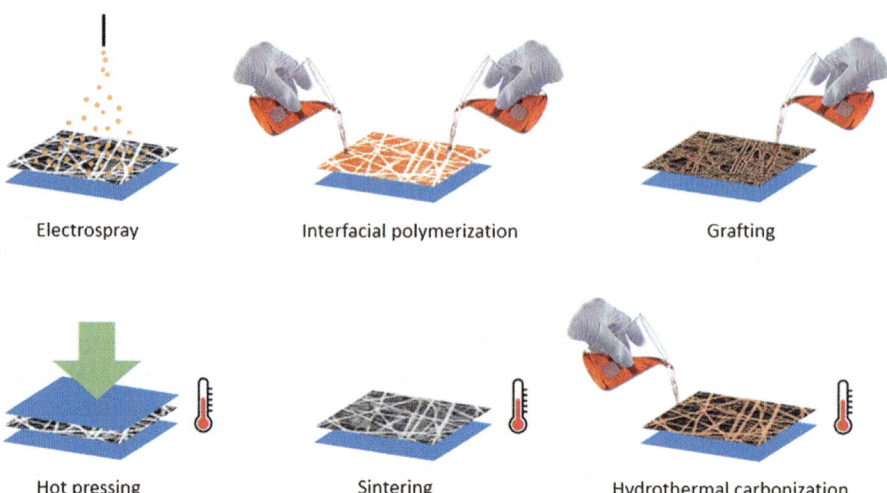

Fig. 1.3 Sketch of different post-treatments for electrospun membranes

chemical warfare agents by simultaneous electrospinning and electrospraying. Electrospinning was carried out feeding a solution of *m*-aramid Nomex dissolved in *N,N*-dimethylacetamide and lithium chloride (1:5 weight ratio of LiCl/Nomex) at a flow rate of 0.15 mL/h and a positive voltage of 20–25 kV. Electrospray was carried out using magnesium oxide or polyoxometalate in methanol at a voltage of +12 kV, working distance of 10 cm, and flow rate of 1.5 mL/h. The collector (a rotating drum) was charged at −5 kV. The samples were heat-treated in air at 300 °C for 10 min to improve their chemical and mechanical properties. The samples were then treated with a fluorinated compound (Unidyne) in order to obtain amphiphobic surfaces. The assemblies simultaneously exhibited super-hydrophobicity (with a water contact angle greater than 150°) and high oleophobicity (oil contact angle higher than 120°). The surface roughness seems to be responsible for the enhanced repellency properties.

In another work [89], the poly(vinylidene fluoride-*co*-hexafluoropropylene) electrospun membrane was treated by electrospray in order to deposit polydimethylsiloxane microspheres with super-hydrophobic properties. Poly(vinylidene fluoride-*co*-hexafluoropropylene) was electrospun from a solution of lithium chloride, *N*-dimethylformamide, and acetone at a voltage of +18 kV, a working distance of 15 cm, and a flow rate of 0.7 mL/h, while microspheres were deposited on the nanofibers by electrospray with a solution of polydimethylsiloxane in *N*-dimethylformamide and tetrahydrofuran solution at a voltage of +30 kV, a working distance 11 cm, and a flow rate of 2.0 mL/h. The resulting hybrid membranes showed a significant increase in roughness and super-hydrophobicity with a contact angle of 155.4°. The membranes provided antifouling properties in the treatment of dyeing wastewaters.

The same strategy was used [90] to coat polyacrylonitrile electrospun nanofibers with silica nanoparticles with the aim of producing super-hydrophilic and underwater super-oleophobic nanofiber membranes for high-flux water–oil separation.

Benefits in using electrospinning and electrospray together were observed also in the production of membranes for carbon dioxide adsorption [91]. This paper reported that a nanofiber membrane was produced by electrospinning of metal oxide nanoparticles dispersed in a polyacrylonitrile solution. The electrospinning was followed by additional metal oxide nanoparticles deposition by electrospray at a high voltage of 40–45 kV with a working distance of 15 cm and a high flow rate of 3.5–4 mL/h. Metal oxide deposition of nanoparticles on the electrospun nanofibers was uniform thanks to electrostatic attraction. The amount of metal oxide was significantly increased, and the carbon dioxide absorption was notably enhanced, as well as the duration of the membrane.

In desalination, one of the most used modifications of nanofibers is by interfacial polymerization because each selective layer can be adjusted independently fulfilling a specific purpose [92].

In a paper [93], polyacrylonitrile nanofibers (with an average diameter of ~500 nm) were produced by electrospinning from a solution in N,N-dimethylacetamide/N,N-dimethylformamide (1:1) at +15 kV voltage, 1.2 mL/h flow rate, and 15 cm working distance. Then a thin polyamide layer was deposited on the nanofibers by interfacial polymerization between m-phenylenediamine and 1,3,5-trimesoyl chloride. The electrospun membrane was first impregnated with a m-phenylenediamine water solution, and second with a 1,3,5-trimesoyl chloride hexane solution. Clearly, water and hexane did not mix and the polymerization happened at the interface. Finally, the freshly prepared membrane was cured under vacuum at 80 °C. The resulting nanofiber membranes are highly hydrophilic and quickly absorb a water droplet. They were tested in osmosis separation of tetracycline hydrochloride in order to simulate antibiotic recovery from water. The membranes showed a significant increase in water flux compared to commercial membranes with an almost complete rejection (99.8%) of tetracycline hydrochloride, indicating great potential for the treatment of antibiotic wastewater.

In another work [94], a silica nanoparticles and polyetherimide mixture was electrospun into nanofibers (with an average diameter of ~250 nm) from a solution of N,N-dimethylformamide/N-methyl-2-pyrrolidone/lithium chloride at +30 kV voltage, 0.9 mL/h flow rate, and 12 cm working distance. A thin polyamide layer was produced on the nanofibers by interfacial polymerization between m-phenylenediamine and 1,3,5-trimesoyl chloride. The nanofibers were first soaked in isopropyl alcohol and washed in water. Then the procedure for the interfacial polymerization is similar to the previous one. The nanofibers were impregnated with an m-phenylenediamine solution in water and with a trimesoyl chloride solution in hexane. Finally, the membrane was cured at 90 °C in water. In this case, the contact angle is close to 130°. The membranes were tested for osmosis separation of a water solution of sodium chloride. The membranes with the highest amount of silica achieved the maximum substrate porosity and the highest water permeability. Moreover, the paper demonstrates that nanocomposite nanofiber membranes resulted in a highly porous structure that facilitates the mass transfer through the layer, and thereby significantly improves the osmotic performance.

Polyethylene terephthalate nanofibers, produced by dual electrospinning from two solutions with different polymer concentrations, were used as support for the production of membranes for osmosis [84]. The nanofibers were heated at 160 °C in order to remove traces of the solvents. Then, interfacial polymerization was carried out using p-phenylenediamine and triethylamine in water and trimesoyl chloride in hexane. Because of the hydrophobic nature of polyethylene terephthalate nanofibers, the organic phase was used first followed by the aqueous phase. The membranes were tested against different salts showing improved performances as compared with a film without nanofibers.

Another technique to reduce the pore size and porosity of electrospun membranes in order to make them suitable for desalination processes is hot pressing. It has been used on electrospun nanofibers to investigate its effect on the mechanical and physical properties of dual-layer nanofiber membranes composed of hydrophobic poly(vinylidene fluoride-co-hexafluoropropylene) at the top layer supported by different hydrophilic nanofiber layers made of polyvinyl alcohol, nylon 6, or polyacrylonitrile [95]. The study had the aim of a better understanding of the role of dual-layer electrospun nanofiber membranes for their performance in desalination. Hydrophobic layer had a water contact angle of 140°, while the other side had a hydrophilic behavior with a water contact angle lower than 90°. Electrospun membranes were dried at 60 °C for 2 days to remove the solvent residues. Hot pressing was then used to press the nanofiber membranes between two flat plates at 170 °C for 1.5 h. The desalination tests were performed for 10 h with 3.5 wt% sodium chloride solution at a temperature of 60 °C. The best performance of rejection greater than 99% has been attributed to the small thickness of the hydrophobic layer.

Sintering is a common method to change physical features of poly(tetrafluoroethylene)-based nanofibers. Huang et al. [81] produced superhydrophobic nanofiber membranes composed of poly(tetrafluoroethylene)/poly(vinyl alcohol) and reinforced with poly(acrylonitrile) nanofibers. Sintering was carried out at different temperatures in order to study its effect on the memmbrane performance. The sintering process fully removed poly(vinyl alcohol) and melted the poly(tetrafluoroethylene). The membranes obtained at different sintering temperatures showed interconnected fiber networks, but the increase in the sintering temperature produced the compactness of the membranes and a reduction of pore size. During membrane distillation tests, salt rejection was greater than 99.8%. Moreover, the high hydrophobicity and the strong lipophilicity would be useful for oil–water separation.

1.4.2 Chemical Modifications

Several strategies were used to enhance hydrophobicity of the nanofibers when required by the operation. For instance, in a membrane for water/oil separation, cellulose acetate on the surface of coaxially electrospun cellulose acetate/polyimide core/shell fibers was modified with fluorinated polybenzoxazine (by in situ polymerization) and silica nanoparticles [74]. Fluorinated surface with silica nanoparticles showed high hydrophobic behavior. Hydrophobicity was evaluated by dynamic contact angle and sliding angle. Advancing and receding contact angles of the membranes were 162.1° and 155.7°; while a water droplet rolled off on a sliding angle of ~7.5°. On the other hand, an oily liquid (dichloromethane) was absorbed by the membrane in less than 150 s. The efficacy to separate oil and water was evaluated in a gravity-driven separation experiment using 20 mL of a mixture dichloromethane and water (50 vol%). Dichloromethane quickly permeated through the membranes, while water was not able to pass through because of the super-hydrophobicity of the membrane.

Grafting was used to functionalize nanofibers for different purposes. Chitpong and Husson [96, 97] chemically modified electrospun regenerated cellulose (i.e., deacetylated cellulose acetate) nanofibers with poly(glycidyl methacrylate). Then, grafting with poly(acrylic acid) or poly(itaconic acid) were carried out in order to produce membranes for heavy metal ions removal from water by ion-exchange processes. The membranes were tested against cadmium, in particular.

Grafting could lead to morphological changes on the nanofiber surfaces, as well as chemical modification. An example was reported by Wang et al. [98]. In the paper, electrospun regenerated cellulose (deacetylated cellulose acetate) nanofibers were chemically modified by atom transfer radical polymerization of 2-hydroxyethyl methacrylate or sodium acrylate. The forming polymer chains, respectively of poly(2-hydroxyethyl methacrylate) and poly(sodium acrylate), were grown on the cellulose surface in order to change pore sizes of electrospun cellulose nanofiber membranes. In particular, poly(2-hydroxyethyl methacrylate) covers the fiber surfaces making the fibers thicker over time and partially filled the pores of the membranes. Poly(sodium acrylate) did not change the fiber size, but it partially filled the pores with a loose polymer structure. Nanoparticles with ~40 nm of size were used to test the membranes for ultrafiltration. After the grafting of poly(2-hydroxyethyl methacrylate), the membranes rejected more than 95% of the nanoparticles, while after the grafting of poly(sodium acrylate) rejected more than 90% of the nanoparticles. The efficiencies were significantly improved compared to unmodified membranes.

Hydrothermal carbonization was used on polyacrylonitrile electrospun nanofibers for adsorption for anionic pollutants from water [99]. Polyacrylonitrile was dissolved in N,N-dimethylformamide and electrospun into nanofibers at +18 kV voltage and at a distance of 15 cm. Hydrothermal carbonization of polyacrylonitrile electrospun nanofibers was carried out on a solution of glucose and diethylenetriamine in water at 160 °C for 18 h. As result of the thermal treatment, the nanofibers were coated with a layer of carbon and amine groups. The membranes were tested as adsorbents of chromium IV ions and 2,4-dichlorophenoxyacetic acid (an herbicide) from water. Adsorption capacities of 290.70 mg/g and 164.47 mg/g for chromium and 2,4-dichlorophenoxyacetic acid were measured, respectively. The adsorption properties were restored by a simple treatment, and the removal efficiencies on both pollutants retained >90% even after five cycles.

1.5 Conclusions and Future Outlooks in Filtering Applications

Electrostatic spinning or electrospinning certainly represents one of the most promising technologies to develop versatile fabric devices to be efficaciously used as smart membranes with ameliorated quality factor with respect to currently commercialized filters. Indeed, it is a promising technology not only for its unequaled operational versatility but also for the unique opportunity to be effectively upscaled, opening actual perspectives for industrial production. At laboratory scale, electrospinning allows the processing of up to several liters of polymer solution under continuous runs, so producing membranes with some milliliters as thickness. Moreover, the intriguing peculiarities of electrospun nanofibers including high surface area and tunable porosity [99], intrinsic three-dimensional (3D) topography and functional properties [100], confer to electrospun membranes great potential to develop innovative filters for industrial applications. Despite the surface area of electrospun polymeric fibers (8–30 m^2/g) being relatively lower than the activated carbon granules, their functionalization by various nanoparticles and/or specific additives may be efficiently used to increase the surface area to capture, thus overcoming their only apparent limiting factor [101]. Innovative structures such as hybrid [102], bead on string [18], and multilayered [103] are concretely demonstrating the possibility to improve the trapping properties of electrospun filters, by different ways. For instance, in the case of hybrid structures, fine fibers contribute to filtration efficiency, while coarser fibers contribute to rigidity. They may be efficiently used to develop integrated filters with hybrid fiber-based sandwich patterns for multipurpose applications [104]. In the case of bead on string structures, obtained by processing via electrospinning low-concentration polymer solution on a continuous matrix of polymeric nanofibers, beads concur to form pores into the fiber network, further improving high filtration efficiency and a relatively low pressure drop. In the case of multilayer composite structures obtained by twin-jet sequential deposition, the heterogeneous nature of membrane, due to the processing of different solutions, promotes the formation of a semi-interpenetrating (i.e., bonding and non-bonding) structure, with variable filtration efficiency, strictly dependent upon the fiber layer arrangement [105, 106]. Despite several studies being reported on the effectiveness of electrospun nanofiber membranes for air and gas filtration, main constraints are ascribable to the large variability of filter design parameters as a function of air pullulans properties in specific environmental contexts, so remarking different spinoffs on human health. In perspective, the main challenge for the electrospun nanofiber membrane will be to design promising filter media to capture differently sized units—that is, nanoparticles and gaseous pollutants—simultaneously. Although electrospun nanofiber membranes are effective in capturing particles of size less than 1000 nm, their potential to capture particles 1–100 nm is not yet established.

In the future, strong interest could be devoted to the implementation of new technologies aimed at improving the filtration through electrostatic attraction. Moreover, it will be necessary to deeply investigate predictive models by the implementation of advanced computational methods able to mathematically optimize process parameters (i.e., fiber diameter, pore size, and nanofiber membrane thickness) in order to empirically achieve maximum quality factor, thus finding the right process-structure-properties correlations for ab initio design filter media.

References

1. Daniel MK, Donald RC, Wenhua HZ, Kenneth CW, Nelms RM, Bruce JT (2007) Fuel cell cathode air filters: methodologies for design and optimization. J Power Sources 168:391–399
2. Kampa M, Castanas E (2008) Human health effects of air pollution. Environ Pollut 151:362–367
3. Dai J, Chen R, Meng X et al (2015) Ambient air pollution, temperature and out-of-hospital coronary deaths in Shanghai, China. Environ Pollut 203:116–121
4. Wang Y, Ying Q, Hu J et al (2014) Spatial and temporal variations of six criteria air pollutants in 31 provincial capital cities in China during 2013–2014. Environ Int 73:413–422
5. Nasreen SAAN, Sundarrajan S, Nizar SAS, Balamurugan R, Ramakrishna S (2013) Advancement in electrospun nanofibrous membranes modification and their application in water treatment. Membranes 3(4):266–284
6. Sanchez JR, Rodriguez JM, Alvaro A et al (1997) Comparative study of different fabrics in the filtration of an aerosol using more complete filtration indexes. Filtr Sep 34:593–598
7. Hung CH, Leung WWF (2011) Filtration of nano-aerosol using nanofibre filter under low Peclet number and transitional flow regime. Sep Purif Technol 79:34–42
8. Wang H, Zheng G, Wang X et al (eds) (2010) Study on the air filtration performance of nanofibrous membranes compared with conventional fibrous filters. In: 5th IEEE international conference on nano/micro engineered and molecular systems (NEMS). IEEE, London, UK, pp 387–390
9. US-EPA (1996) Air quality criteria for particulate matter. US Environmental Protection Agency, EPA/600/P-95/001F, Washington, DC
10. Brown RC (2001) Filtration in industrial hygiene. Am Ind Hyg Assoc J 62:633–643
11. Qiu H, Tian L, Ho K et al (2015) Air pollution and mortality: effect modification by personal characteristics and specific cause of death in a case-only study. Environ Pollut 199:192–197
12. Thavasi V, Singh G, Ramakrishna S (2008) Electrospun nanofibres in energy and environmental applications. Energy Environ Sci 1:205–221
13. Wang N, Raza A, Si Y et al (2013) Tortuously structured polyvinyl chloride/polyurethane fibrous membranes for high-efficiency fine particulate filtration. J Colloid Interface Sci 398:240–246
14. Gorji M, Jeddi A, Gharehaghaji A (2012) Fabrication and characterization of polyurethane electrospun nanofibre membranes for protective clothing applications. J Appl Polym Sci 125:4135–4141
15. Schreuder-Gibson H, Gibson P et al (2002) Protective textile materials based on electrospun nanofibres. J Adv Mater 34:44–55
16. Desai K, Kit K, Li J et al (2009) Nanofibrous chitosan non-wovens for filtration applications. Polymer 50:3661–3669
17. Aluigi A, Vineis C, Tonin C et al (2009) Wool keratin-based nanofibres for active filtration of air and water. J Biobased Mater Bioenergy 3:311–319
18. Wang Z, Zhao C, Pan Z (2015) Porous bead-on-string poly(lactic acid) fibrous membranes for air filtration. J Colloid Interface Sci 441:121–129
19. Han DH, Kang MS (2009) A survey of respirators usage for airborne chemicals in Korea. Ind Health 47:569–577

20. Huang X, Wang YJ, Di YH (2007) Experimental study of wool fibre on purification of indoor air. Text Res J 77:946–950
21. Yao WL, Wang JL, Yang J, Du GD (2008) Novel carbon nanofibre-cobalt oxide composites for lithium storage with large capacity and high reversibility. J Power Sources 176(1):369–372
22. Ju YW, Park JH, Jung HR, Cho SJ, Lee WJ (2008) Electrospun MnFe2O4 nanofibres: preparation and morphology. Compos Sci Technol 68(7–8):1704–1709
23. Park SH, Kim C, Choi YO, Yang KS (2003) Preparations of pitch-based CF/ACF webs by electrospinning. Carbon 41(13):2655–2657
24. Ahn C, Park SK, Kim GT et al (2006) Development of high efficiency nanofilters made of nanofibres. Curr Appl Phys 6(6):1030–1035
25. Reneker DH, Chun I (1996) Nanometre diameter fibres of polymer, produced by electrospinning. Nanotechnology 7:216
26. Deitzel JM, Kleinmeyer J, Harris D, Beck Tan NC (2001) The effect of processing variables on the morphology of electrospun nanofibers and textiles. Polymer 42:261
27. Yarin AL, Koombhongse S, Reneker DH (2001) Bending instability in electrospinning of nanofibers. J Appl Phys 89(5):3018
28. Greiner A, Wendorff JH (2007) Electrospinning: a fascinating method for the preparation of ultrathin fibers. Angew Chem Int Ed 46:5670
29. Sill TJ, von Recum HA (2008) Electrospinning: applications in drug delivery and tissue engineering. Biomaterials 29(13):1989–2006
30. Guarino V, Cirillo V, Taddei P, Alvarez-Perez MA, Ambrosio L (2011) Tuning size scale and cristalliniy of PCL electrospun membranes via solvent permittivity to address hMSC response. Macromol Biosci 11:1694–1705
31. Megelski S, Stephens JS, Chase DB, Rabolt JF (2002) Micro- and nanostructured surface morphology on electrospun polymer fibres. Macromolecules 35(22):8456–8466
32. Kumar A, Wei M, Barry C, Chen J, Mead J (2010) Controlling fibre repulsion in multi-jet electrospinning for higher throughput. Macromol Mater Eng 295:701–708. https://doi.org/10.1002/mame.200900425
33. Eda G, Shivkumar S (2006) Bead structure variations during electrospinning of polystyrene. J Mater Sci 41(17):5704–5708
34. Buchko CJ, Chen LC, Shen Y, Martin DC (1999) Processing and microstructural characterization of porous biocompatible protein polymer thin films. Polymer 40(26):7397–7407
35. Guarino V, Albore MD, Altobelli R, Ambrosio L (2016) Polymer bioprocessing to fabricate 3D scaffolds for tissue engineering. Int Polym Process 31:587. https://doi.org/10.3139/217.3239
36. Jiang S, Hou H, Agarwal S, Greiner A (2016) Polyimide nanofibres by "green" electrospinning via aqueous solution for filtration applications. ACS Sustain Chem Eng 4(9):4797–4804
37. Ko J, Mohtaram NK, Ahmed F, Montgomery A, Carlson M, Lee PC, Willerth SM, Jun MB (2014) Fabrication of Poly(ecaprolactone) microfibre scaffolds with varying topography and mechanical properties for stem cell-based tissue engineering applications. J Biomater Sci Polym 25:1–17
38. Guarino V, Ambrosio L (2016) Electrofluidodynamics: exploring new toolbox to design biomaterials for tissue regeneration and degeneration. Nanomedicine 11(12):1515–1518
39. Cirillo V, Guarino V, Alvarez-Perez MA, Marrese M, Ambrosio L (2014) Optimization of fully aligned bioactive electrospun fibres for "in vitro" nerve guidance. J Mater Sci Mater Med 25(10):2323–2332
40. Guarino V, Cirillo V, Ambrosio L (2016) Bicomponent electrospun scaffolds to design ECM tissue analogues. Exp Rev Med Dev 13(1):83–102
41. Detta N, Brown TD, Edin FK, Albrecht K, Chiellini F, Chiellini E, Dalton PD, Hutmacher DW (2010) Melt electrospinning of polycaprolactone and its blends with poly (ethylene glycol). Polym Int 59:1558–1562
42. Patil H, Tiwari RV, Repka MA (2016) Hot-melt extrusion: from theory to application in pharmaceutical formulation. AAPS PharmSciTech 17:20–42

43. Schmack G, Tändler B, Vogel R, Beyreuther R, Jacobens S, Fritz HG (1999) Biodegradable fibres of Poly(l-lactide) produced by high-speed melt spinning and spin drawing. J Appl Polym Sci 73:2785–2797

44. Muerza-Cascante ML, Haylock D, Hutmacher DW, Dalton PD (2015) Melt electrospinning and its technologization in tissue engineering. Tissue Eng Part B 21:187–202

45. Farrugia BL, Brown TD, Upton Z, Hutmacher DW, Dalton PD, Dargaville TR (2013) Dermal fibroblast infiltration of Poly(e-caprolactone) scaffolds fabricated by melt electrospinning in a direct writing mode. Biofabrication 5:025001

46. Qin X-H, Wang S-Y (2008) Electrospun nanofibers from cross-linked poly(vinyl alcohol) and its filtration efficiency. J Appl Polym Sci 109(2):951–956

47. Homaeigohar SS, Buhr K, Ebert K (2010) Polyethersulfone electrospun nanofibrous composite membrane for liquid filtration. J Membr Sci 365(1–2):68–77

48. Polat Y, Pampal ES, Stojanovska E, Simsek R, Hassanin A, Kilic A et al (2016) Solution blowing of thermoplastic polyurethane nanofibers: a facile method to produce flexible porous materials. J Appl Polym Sci 133(9):43025–43034

49. Medeiros ES, Glenn GM, Klamczynski AP, Orts WJ, Mattoso LHC (2009) Solution blow spinning: a new method to produce micro- and nanofibres from polymer solutions. J Appl Polym Sci 113(4):2322–2330

50. Sinha-Ray S, Sinha-Ray S, Yarin AL, Pourdeyhimi B (2015) Theoretical and experimental investigation of physical mechanisms responsible for polymer nanofibre formation in solution blowing. Polymer 56:452–463

51. Zhuang X, Yang X, Shi L, Cheng B, Guan K, Kang W (2012) Solution blowing of submicron-scale cellulose fibres. Carbohydr Polym 90(2):982–987

52. Tao X, Zhou G, Zhuang X, Cheng B, Li X, Li H (2015) Solution blowing of activated carbon nanofibres for phenol adsorption. RSC Adv 5(8):5801–5808

53. Hsiao H-Y, Huang C-M, Hsu M-Y, Chen H (2011) Preparation of high-surface-area PAN-based activated carbon by solution-blowing process for CO2 adsorption. Sep Purif Technol 82:19–27

54. Ding B, Kimura E, Sato T, Fujita S, Shiratori S (2004) Fabrication of blend biodegradable nanofibrous nonwoven mats via multi-jet electrospinning. Polymer 45:1895

55. Teo WE, Ramakrishna S (2005) Electrospun fibre bundle made of aligned nanofibres over two fixed points. Nanotechnology 16:1878

56. Kim G, Cho YS, Kim WD (2006) Stability analysis for multi-jets electrospinning process modified with a cylindrical electrode. Eur Polym J 42:2031

57. Varabhas JS, Chase GG, Reneker DH (2008) Electrospun nanofibers from a porous hollow tube. Polymer 49:4226

58. Zhou F-L, Gong R-H, Porat I (2010) Needle and needleless electrospinning for nanofibres. J Appl Polym Sci 115:2591–2598. https://doi.org/10.1002/app.31282

59. Niu H, Lin T (2012) Fibre generators in needleless electrospinning. J Nanomater 2012:725950

60. Moon S, Gil M, Lee KJ (2017) Syringeless electrospinning toward versatile fabrication of nanofibre web. Sci Rep 7:41424

61. Niu H, Wang X, Lin T (2011) Needleless electrospinning: developments and performances, nanofibres – production, properties and functional applications, Dr. Lin T (ed). ISBN: 978-953-307-420-7

62. Teo WE, Ramakrishna S (2006) A review on electrospinning design and nanofibre assemblies. Nanotechnology 17(14):R89

63. Kongkhlang T, Tashiro K, Kotaki M, Chirachanchai S (2008) Electrospinning as a new technique to control the crystal morphology and molecular orientation of polyoxymethylene nanofibers. J Am Chem Soc 130(46):15460–15466

64. Katta P, Alessandro M, Ramsier RD, Chase GG (2004) Continuous electrospinning of aligned polymer nanofibers onto a wire drum collector. Nano Lett 4(11):2215–2218

65. Huang ZM, Zhang YZ, Kotaki M, Ramakrishna S (2003) A review on polymer nanofibers by electrospinning and their applications in nanocomposites. Compos Sci Technol 63(15):2223–2253. 26–30

66. Yee WA, Nguyen AC, Lee PS, Kotaki M, Liu Y, Tan BT, Mhaisalkar S, Lu X (2008) Stress-induced structural changes in electrospun polyvinylidene difluoride nanofibers collected using a modified rotating disk. Polymer 49(19):4196–4203

67. Kim K, Lee K, Khil M, Ho Y, Kim H (2004) The effect of molecular weight and the linear velocity of drum surface on the properties of electrospun poly(ethylene terephthalate) non-wovens. Fibres Polym 5(2):122–127

68. Wang Y, Wang G, Chen L, Li H, Yin T, Wang B, Lee JCM, Yu Q (2009) Electrospun nanofibre meshes with tailored architectures and patterns as potential tissue-engineering scaffolds. Biofabrication 1:015001:1–015001:9

69. Li D, Wang YL, Xia YN (2004) Electrospinning nanofibres as uniaxially aligned arrays and layer-by-layer stacked films. Adv Mater 16:361–366

70. Zussman E, Theron A, Yarin AL (2003) Formation of nanofibre crossbars in electrospinning. Appl Phys Lett 82:973–975

71. Ye XY, Huang XJ, Xu ZK (2012) Nanofibrous mats with bird's nest patterns by electrospinning. Chin J Polym Sci 30:130–137

72. Ding Z, Salim A, Ziaie B (2009) Selective nanofibre deposition through field-enhanced electrospinning. Langmuir 25:9648–9652

73. Zucchelli A, Fabiani D, Gualandi C, Focarete ML (2009) An innovative and versatile approach to design highly porous, patterned, nanofibrous polymeric materials. J Mater Sci 44:4969–4975

74. Ma W, Guo Z, Zhao J, Yu Q, Wang F, Han J, Pan H, Yao J, Zhang Q, Samal SK, De Smedt SC, Huang C (2017) Polyimide/cellulose acetate core/shell electrospun fibrous membranes for oil-water separation. Sep Purif Technol 177:71–85

75. Lee E-J, An AK, Hadi P, Lee S, Wood YC, Shon HK (2017) Advanced multi-nozzle electrospun functionalized titanium dioxide/polyvinylidene fluoride-co-hexafluoropropylene (TiO$_2$/PVDF-HFP) composite membranes for direct contact membrane distillation. J Membr Sci 524:712–720

76. Sun H, Xu Y, Zhou Y, Gao W, Zhao H, Wang W (2017) Preparation of superhydrophobic nanocomposite fibre membranes by electrospinning poly(vinylidene fluoride)/silane coupling agent modified SiO$_2$ nanoparticles. J Appl Polym Sci 134:44501

77. Wei Z, Li J, Wang C, Cao J, Yao Y, Lu H, Li Y, He X (2017) Thermally stable hydrophobicity in electrospun silica/polydimethylsiloxane hybrid fibres. Appl Surf Sci 392:260–267

78. Woo YC, Tijing LD, Shim W-G, Choi J-S, Kim S-H, He T, Drioli E, Shon HK (2016) Water desalination using graphene-enhanced electrospun nanofibre membrane via air gap membrane distillation. J Membr Sci 520:99–110

79. Ren L-F, Xia F, Shao J, Zhang X, Li J (2017) Experimental investigation of the effect of electrospinning parameters on properties of superhydrophobic PDMS/PMMA membrane and its application in membrane distillation. Desalination 404:155–166

80. Cojocaru C, Dorneanu PP, Airinei A, Olaru N, Samoila P, Rotaru A (2017) Design and evaluation of electrospun polysulfone fibres and polysulfone/NiFe$_2$O$_4$ nanostructured composite as sorbents for oil spill cleanup. J Taiwan Inst Chem Eng 70:267–281

81. Huang Y, Huang Q-L, Liu H, Zhang C-X, You Y-W, Li N-N, Xiao C-F (2017) Preparation, characterization, and applications of electrospun ultrafine fibrous PTFE porous membranes. J Membr Sci 523:317–326

82. Pan J, Hou L, Wang Q, He Y, Wu L, Mondal AN, Xu T (2017) Preparation of bipolar membranes by electrospinning. Mater Chem Phys 186:484–491

83. Wang J, Zhao W, Wang B, Pei G, Li C (2017) Multilevel-layer-structured polyamide/poly(trimethylene terephthalate) nanofibrous membranes for low-pressure air filtration. J Appl Polym Sci 134:44716

84. Mahdavi H, Moslehi M (2016) A new thin film composite nanofiltration membrane based on PET nanofibre support and polyamide top layer: preparation and characterization. J Polym Res 23:257
85. Lou L-H, Qin X-H, Zhang H (2017) Preparation and study of low-resistance polyacrylonitrile nano membranes for gas filtration. Text Res J 87:208–215
86. Zhang S, Liu H, Yin X, Li Z, Yu J, Ding B (2017) Tailoring mechanically robust Poly(m-phenylene isophthalamide) nanofibre/nets for ultrathin high-efficiency air filter. Sci Rep 7:40550
87. Aslan T, Arslan S, Eyvaz M, Güçlü S, Yüksel E, Koyuncu İ (2016) A novel nanofibre microfiltration membrane: Fabrication and characterization of tubular electrospun nanofibre (TuEN) membrane. J Membr Sci 520:616–629
88. Ryu S-Y, Chung JW, Kwak S-Y (2017) Tunable multilayer assemblies of nanofibrous composite mats as permeable protective materials against chemical warfare agents. RSC Adv 7:9964–9974
89. An AK, Guo J, Lee E-J, Jeong S, Zhao Y, Wang Z, Leiknes T (2017) PDMS/PVDF hybrid electrospun membrane with superhydrophobic property and drop impact dynamics for dyeing wastewater treatment using membrane distillation. J Membr Sci 525:57–67
90. Ge J, Zhang J, Wang F, Li Z, Yu J, Ding B (2017) Superhydrophilic and underwater superoleophobic nanofibrous membrane with hierarchical structured skin for effective oil-in-water emulsion separation. J Mater Chem A 5:497–502
91. Wahiduzzaman, Allmond K, Stone J, Harp S, Mujibur K (2017) Synthesis and electrospraying of nanoscale MOF (metal organic framework) for high-performance CO2 adsorption membrane. Nanoscale Res Lett 12:6
92. Chowdhury MR, Huang L, McCutcheon JR (2017) Thin film composite membranes for forward osmosis supported by commercial nanofibre nonwovens. Ind Eng Chem Res 56:1057–1063
93. Pana S-F, Dong Y, Zhenga Y-M, Zhong L-B, Yuan Z-H (2017) Self-sustained hydrophilic nanofibre thin film composite forward osmosis membranes: preparation, characterization and application for simulated antibiotic wastewater treatment. J Membr Sci 523:205–215
94. Tian M, Wang Y-N, Wang R, Fane AG (2017) Synthesis and characterization of thin film nanocomposite forward osmosis membranes supported by silica nanoparticle incorporated nanofibrous substrate. Desalination 401:142–150
95. Woo YC, Tijing LD, Park MJ, Yao M, Choi J-S, Lee S, Kim S-H, An K-J, Shon HK (2017) Electrospun dual-layer nonwoven membrane for desalination by air gap membrane distillation. Desalination 403:187–198
96. Chitpong N, Husson SM (2016) Nanofibre ion-exchange membranes for the rapid uptake and recovery of heavy metals from water. Membranes 6:59
97. Chitpong N, Husson SM (2017) Polyacid functionalized cellulose nanofibre membranes for removal of heavy metals from impaired waters. J Membr Sci 523:418–429
98. Wang Z, Crandall C, Prautzsch VL, Sahadevan R, Menkhaus TJ, Fong H (2017) Electrospun regenerated cellulose nanofibre membranes surface-grafted with water-insoluble poly(HEMA) or water-soluble poly(AAS) chains via the ATRP method for ultrafiltration of water. ACS Appl Mater Interfaces 9:4272–4278
99. Welle A, Kröger M, Döring M, Niederer K, Pindel E, Chronakis IS (2007) Electrospun aliphatic polycarbonates as tailored tissue scaffold materials. Biomaterials 28:2211
100. Dhandayuthapani B, Poulose AC, Nagaoka Y, Hasumura T, Yoshida Y, Maekawa T, Kumar DS (2012) Biomimetic smart nanocomposite: in vitro biological evaluation of zein electrospun fluorescent nanofiber encapsulated CdS quantum dots. Biofabrication 4:025008
101. Kim HJ, Pant HR, Choi NJ et al (2013) Composite electrospun fly ash/polyurethane fibres for absorption of volatile organic compounds from air. Chem Eng J 230:244–250
102. Yarin AL, Pourdeyhimi B, Ramakrishna S (2014) Fundamentals and applications of micro and nanofibres. Cambridge University Press, Cambridge

103. Wang N, Si Y, Wang N et al (2014) Multilevel structured polyacrylonitrile/silica nanofibrous membranes for high-performance air filtration. Sep Purif Technol 126:44–51
104. Sutherland K (2013) Air filtration: keeping the air that we breathe clean. Filtr Sep 50:18–22
105. Wang N, Yang Y, Al-Deyab SS et al (2015) Ultra-light 3D nanofibre-nets binary structured nylon 6–polyacrylonitrile membranes for efficient filtration of fine particulate matter. J Mater Chem A 3:23946–23954
106. Shen J, Chen R, He J (2014) Bio-mimic design of PM2.5 anti-smog masks. Therm Sci 18:1689–1690

Chapter 2
Current Advances on Nanofiber Membranes for Water Purification Applications

Hongyang Ma and Benjamin S. Hsiao

Abstract Electrospun nanofiber membranes have many application potentials, including air/water filtration, gas storage, sensors/electronics, and healthcare/cosmetics. The recent advances of these membranes for water filtration applications, including microfiltration, ultrafiltration, nanofiltration, reverse osmosis, forward osmosis, and membrane distillation, are reviewed here. The high porosity, adjustable pore size/pore size distribution, large range of materials choice, and available surface functionalization have provided the flexibility to tailor-design the membranes for numerous existing and emerging applications. Recent advances in electrospinning technology have further offered a variety of pathway for scale-up production of electrospun membranes, realizing its potential for water purification.

2.1 Introduction

Quasi-three-dimensional nanofibrous scaffolds with submicron or nano-sized diameters can be fabricated using the electrospinning technology [1–4]. Electrospinning is the atomization process of a conducting fluid under high electrostatic field. The fluid is usually a charged polymer solution, where the solution jet can be ejected toward the target collector when the applied electrostatic field overcomes the surface tension of the polymer solution, as shown in Fig. 2.1a [5]. The fluid jet first travels in a straight line, followed by bending and twisting due to solvent evaporation, resulting in a significant increase of electrostatic repulsion, as shown in

H. Ma (✉)
State Key Laboratory of Organic-Inorganic Composites,
Beijing University of Chemical Technology, Beijing, China

Department of Chemistry, Stony Brook University, Stony Brook, NY, USA
e-mail: mahy@mail.buct.edu.cn

B. S. Hsiao (✉)
Department of Chemistry, Stony Brook University, Stony Brook, NY, USA
e-mail: Benjamin.Hsiao@stonybrook.edu

© Springer International Publishing AG, part of Springer Nature 2018
M. L. Focarete et al. (eds.), *Filtering Media by Electrospinning*,
https://doi.org/10.1007/978-3-319-78163-1_2

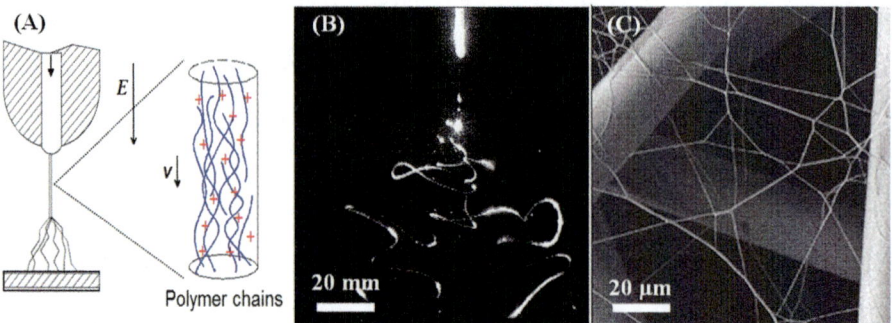

Fig. 2.1 Polymeric fluid droplet under an applied electric field (**a**) [5], high-speed photograph of electrospinning process (**b**) Image courtesy of Dr. Christian Burger, Stony Brook University [6], and SEM image of PET with electrospun nanofibers (**c**) [7]. Reprinted with permission from reference [5]. Copyright (2006) American Chemical Society

Fig. 2.1b [6]. The unstable jet leads to a substantial stretching of the polymer chains, and forms a nanoscale size filament when deposited on the collector. With the appropriate control of solvent evaporation, the final filament can contain a small amount of solvent and form an interconnected quasi-three-dimensional mat with all the junctions fused. The diameter of electrospun fiber is often about two orders of magnitude smaller than that of synthetic fibers, as shown in Fig. 2.1c [7].

The electrospun nanofibrous scaffolds with nanoscale fiber diameters have very large surface-area-to-volume ratio and very high porosity [8, 9]. More than that, the morphology and properties of the electrospun nanofibers could be designed and fabricated by using a wide variety of materials (e.g., inorganic or organic polymers, including natural and synthetic, and hybrid materials) and by adjusting the parameters of the polymer solution (e.g., concentration, viscosity, conductivity, polymer type, solvent), electrospinning process (e.g., applied voltage, flow rate, jet travelling distance, collection target materials), and environmental variables (e.g., temperature, humidity, chamber pressure) [2]. It should be mentioned that most of the parameters are closely correlated with one another in the electrospinning process. Consequently, electrospinning is both complex and versatile. Combining the physical features of nanofibers with chemical functionality, electrospun fibrous scaffolds can be used for a broad range of applications, including air and water purification [10, 11], drug (gene) delivery [12], wound dressing [13–15], enzyme immobilization [16], tissue engineering [17–19], gas storage [20], and sensors and electrodes in electronics [21, 22].

In the case of water purification, electrospun nanofibrous scaffolds can be used directly as a barrier layer to remove large contaminants in microfiltration, or indirectly as a substrate layer to support a barrier layer for ultrafiltration, nanofiltration, and reverse osmosis for desalination [11]. There are a number of comprehensive reviews dealing with the subject of electrospun nanofibrous membranes for water purification, where each review focused on the different aspects of the topic [6, 11, 23–26]. For example, one review emphasized the use of a variety of natural and synthetic polymer materials in fabrication of electrospun nanofibrous mats [11], whereas the others focused on the surface modifications schemes [6, 27]. It is clear that the electrospinning technology can offer unique advantages, either through the electrospinning setup or by more flexible materials selection, over conventional methods to design and fabricate membranes with suitable physical and chemical properties to meet some special requirements for a targeted water treatment. Thus, in this chapter, we focus on the recent advances in electrospinning technologies, which are particularly suited to produce uniform nanofibrous scaffolds appropriate for a wide range of water purification applications.

2.2 Electrospinning Technology

2.2.1 Brief History

The electrospinning phenomenon was first observed by Rayleigh in 1879 [28], which led to the seminal demonstration of electrospraying by Zeleny in 1914 [29]. Zeleny carefully investigated the behavior of fluid droplets at the tip of metal capillaries under electrical field using a mathematical model to describe the behavior of charged fluid by electrostatic force. The setup and the method of electrospinning were patented separately by Cooley and Morton, respectively, in 1902 [30, 31]. Further developments toward commercialization were made by Formhals, who filed a sequence of patents in 1934 (e.g., the fabrication of textile yarns [32]). Electrospinning from a melt rather than from a solution was first patented by Norton in 1936, using an air-blast to assist the fiber formation [33]. Thereafter, much research has focused on more fundamental research, especially the formation of the jet from the spinneret as a function of electrostatic field strength, fluid viscosity, and molecular weight of the polymer in solution. Among these studies, the works of Taylor in 1969 on the electrically driven jets were particularly noteworthy, which had been regarded as the foundation for the current technology [34, 35]. Up to now, there are more than 1000 patents and 10,000 research publications on the subject of electrospinning of polymer melts and solutions. The rapid growth of this technology can be appreciated by the fact that, in 2013, a review indicated that there were only about 2000 publications associated with the electrospinning technology, where about 1/10 was related to biomedical applications [36].

2.2.2 Single-Jet Electrospinning

One reason for the rapid spread of the electrospinning technology is the simplicity of the setup. A single-jet electrospinning device can be constructed easily for laboratory research. In a typical electrospinning setup (as illustrated in Fig. 2.2), the major components include a high-voltage power supply, a syringe pump for

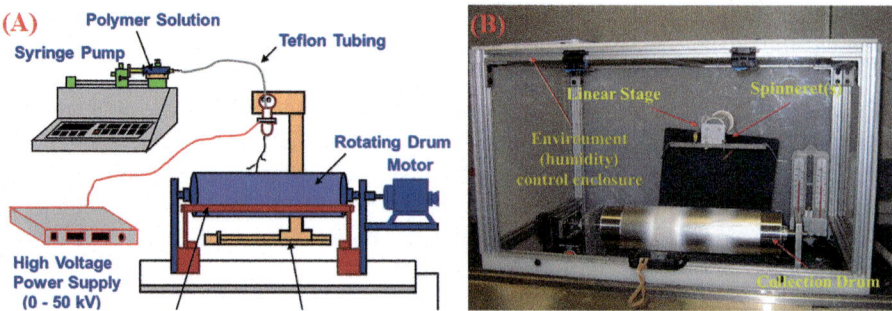

Fig. 2.2 Typical setup of a laboratory electrospinning apparatus (**a**) and a photograph of an enclosure setup for environmental control (**b**) [37]. Reprinted from reference [37], Copyright (2002), with permission from Elsevier

delivery of polymer solution, a spinneret, a collecting drum, and an environmentally controlled enclosure that can adjust environmental parameters, such as temperature, humidity, gas (air) circulation, and solvent evaporation rate [2].

Most of natural materials, such as polysaccharides and silk, and synthetic polymers, such as polyacrylonitrile (PAN), polyethersulfone (PES), and polyvinylidene fluoride (PVDF), often used in water purification, can be fabricated directly into nanofibers using the above setup without an environmental control chamber [38]. However, a chamber with controlled humidity and temperature is often used to ensure the production of high quality nanofiber scaffolds. It is also imperative to point out that a proper housing assembly can provide the necessary protection for potential electrical hazards [39].

For water applications, PAN, PES, PVDF, polyvinyl alcohol (PVA), cellulose acetate (CA) and their derivatives have been electrospun into submicron fibrous scaffolds, which were used either as a barrier layer or as a supporting layer in filtration membranes [40]. The fiber diameter of electrospun nanofibers ranged typically from 0.1 to 1.0 μm, while the high porosity of the scaffolds varied from 70% to 95% [36]. The mean pore size of the nanofibrous scaffold could be adjusted for the removal of nanoparticles or waterborne bacteria (>0.2 μm) from contaminated water by size exclusion. An empirical relationship has been found between the mean pore size and mean fiber diameter of an optimized electrospun nanofibrous membrane for water purification—the mean pore size is about three times the mean fiber diameter when the porosity is fixed at 80% [23]. Therefore, the pore size of the filtration membrane can be designed by controlling the fiber diameter. It is apparent that the interconnected porous structure of an electrospun scaffold, defined by random deposition of nanofibers, can definitely lower the hydraulic resistance of water transportation when compared to conventional porous membranes, resulting in high permeation flux, which is beneficial for both barrier layer and support layer of a membrane [25].

In our recent studies, electrospun PAN membranes have been fabricated to remove bacteria, such as *E. coli* and *B. diminuta*, exhibiting six-log reduction value

(LRV) efficiency, meeting the requirements for drinking water purification [41, 42]. The permeation flux of these electrospun PAN membranes was found to be about 2–5 times higher than those of most commercially available counterparts due to nanofibrous membrane's high porosity and highly interconnected pores. The high flux performance allows more efficient usage in low-pressure driven processes such as gravity-powered filtration. For example, the pressure drop in a gravity-driven filtration process using electrospun nanofibrous membranes could be remained as low as 0.2 psi created using a feed solution containing model contaminants [43].

By infusing ultrafine cellulose nanofibers (CN) into an electrospun PAN scaffold, the composite membrane can further be used to remove waterborne viruses exhibiting 2–4 LRV against *MS2*, a bacteriophage [44]. The infused cellulose nanofibers, having functional groups such as carboxylate, hydroxyl, and aldehyde on the cellulose surface, could form a nanoscale web in the nanofibrous network of larger fiber size and behave as an adsorption medium [45]. This composite membrane structure offers several advantages over conventional microfiltration membranes: (1) a significant increase in the surface-to-volume ratio; (2) a decrease in pore size with reduced pore size distribution by forming a 3-D nanoweb structure among the electrospun nanofibrous scaffold; (3) enhanced overall mechanical properties of the composite network; and (4) abundant functionalities on the cellulose nanofiber surface can serve as adsorption sites to remove contaminants, such as viruses, dyes, heavy metal ions, and toxins [23]. Meanwhile, the high permeation flux was maintained because of the higher porosity, when compared with commercially available membranes at the same operating pressure.

It is clear that the pathway of surface functionalization can provide new opportunities to improve the separation efficiency of electrospun nanofibrous membrane, not just as a microfiltration filter but also as an adsorbent, for water purification. For example, using a coating method, the fiber surface in electrospun PAN membrane could become positively charged, where the resulting membrane can remove both nanoparticle through filtration and BSA protein through adsorption in simulated contaminated water [41]. In specific, electrospun PAN membrane was dip-coated with dual-vinyl monomers, where free radical polymerization could be initialed by a thermal treatment on the nanofiber surface [41]. The cross-linked polymer coating layer, containing imidazolium cations, not only enhanced the mechanical properties (by binding the junction points between the nanofibers) but also could adsorb negatively charged BSA proteins by electrostatic interactions. In the abovementioned scenarios, electrospun nanofibers were used directly as a barrier layer for microfiltration or an adsorbent. On the other hand, electrospun nanofibrous scaffold can also be used as a high porous substrate to support a barrier layer for ultrafiltration [46–54], nanofiltration [55–57], and reverse/forward osmosis applications [58, 59]. The new structure is termed thin-film nanofibrous composite (TFNC) structure, which often consists of three layers: the bottom layer is a conventional nonwoven substrate (e.g., polyethylene terephthalate, PET nonwoven mat); the middle layer is an electrospun nanofibrous scaffold; and the top barrier can be a hydrophilic polymer coating layer or another finer nanofiber layer [48], as shown in Fig. 2.3.

Fig. 2.3 Three-layered nanofibrous membrane structure having cellulose nanofiber barrier layer (fiber diameter about 5 nm), electrospun nanofiber substrate (fiber diameter about 100 nm), and nonwoven fiber support (fiber diameter about 20 μm). Reprinted with permission from reference [49], Copyright (2011) American Chemical Society

Using the single-jet electrospinning technique, several TFNC membrane systems have been demonstrated. In one system, electrospun PAN scaffold was used to support a cross-linked PVA (using glutaraldehyde, GA, as the cross-linking reagent) barrier layer [46–49]. As both bulk and surface porosities of electrospun PAN scaffold (~80%) were significantly higher than those of porous support (bulk porosity ~ 50% and surface porosity ~ 25%) prepared by the phase-inversion method, the permeation flux of the TENC membrane was about 2–10× higher than the conventional membranes for ultrafiltration applications due to the lower hydraulic resistance of the supporting scaffold [25]. In addition to cross-linked PVA [46, 47, 53], other hydrophilic barrier layer using chitosan [60], cellulose [49, 51], and cellulose nanofibers [48, 52, 54] were also demonstrated in TFNC membranes for ultrafiltration applications, which will be discussed further later. The hydrophilic nature of these membranes exhibited antifouling properties, as well as higher permeability [11].

For nanofiltration and reverse osmosis, interfacial polymerization is often used to create a dense barrier layer on the top of an electrospun PAN membrane using the TFNC format [55–59]. In addition to higher permeability, electrospun nanofibrous scaffolds can also reduce the surface polarization of seed solution due to the interconnected porous structure [59]. In one study, up to two times permeation flux was achieved by using electrospun nanofibrous membrane as the mid-layer to support an interfacially polymerized barrier layer containing molecular additives, when compared with commercially available NF membranes (e.g., NF270) [57].

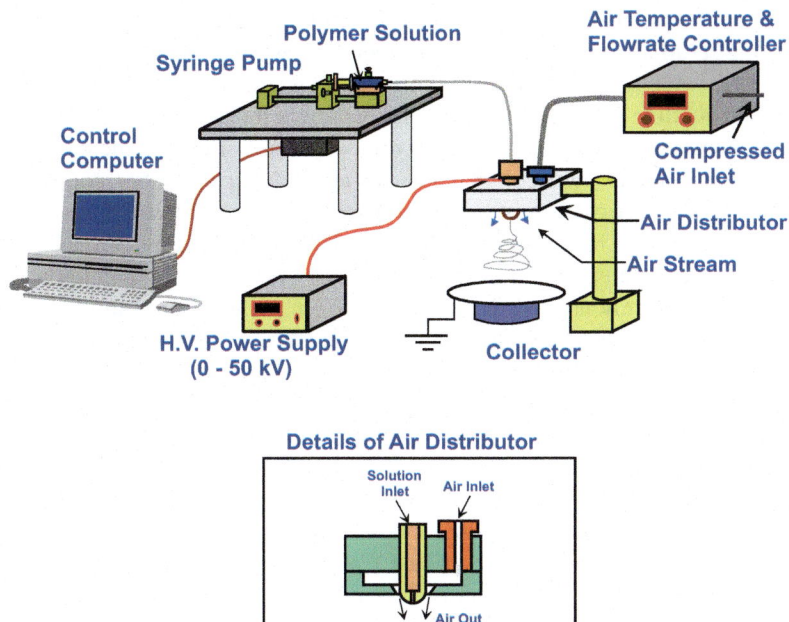

Fig. 2.4 Schematic diagram of electroblowing setup and structural details of the air distributor [64]. Reprinted with permission from reference [64], Copyright (2004) American Chemical Society

Using the single-jet electrospinning technology, electrospun membranes were produced for the membrane distillation application. Membrane distillation (MD) is an alternative approach that can desalinate seawater with significant energy saving benefits. The MD process is a thermal-gradient driven process, where electrospun nanofibrous membrane plays an essential role for separation of pure water and contaminants such as salt ions [61–63]. The ideal membranes for the MD application should have a small pore size (<0.2 μm) and a hydrophobic nature, where this structure and property can allow water vapor to efficiently pass through but prevent the passage of the liquid phase. With high porosity (but small pore size), an electrospun hydrophobic polystyrene (PS) membrane exhibited relatively high mass transfer rate for desalination of seawater and near 99.99% retention against sodium chloride [63].

2.2.3 Electroblowing Technology

A unique electrospinning operation, termed electroblowing, has been demonstrated to fabricate some more difficult-to-process materials, such as hyaluronic acid, into nanofibrous membrane having fibers with diameters in the tens of nanometer size range [64, 65]. We envision this technology can also be used to produce nanofibers as filter or adsorption media for water purification. A schematic diagram of the electroblowing setup for making nanofibrous membranes is shown in Fig. 2.4.

The difference between electrospinning and electroblowing is the adoption of airflow surrounding the spinneret in the latter, where the use of air can increase the production speed and control the solvent removal rate. In Fig. 2.4, the additional air blow system contains two components: an air-blowing assembly and a heating assembly. The gaseous flow rate was regulated directly by a speed-controlled blower, while the air temperature was controlled by the heating element. The air temperatures at different locations of the air blow system, depending upon the airflow rate, could be monitored to adjust the air temperature surrounding the spinneret. With this setup, we have successfully tested the electroblowing processing of some very viscous systems, such as hyaluronan (HA) solutions of different compositions and molecular weights that are difficult to electrospin, under typical electrospinning conditions. A consistent production of electrospun HA nanofibers with fiber diameter of about 100 nm have been fabricated using 2.5 (w/v)% HA with high molecular weight of ~3.5 million Dalton aqueous solution. The as-prepared electrospun HA nanofibers could be further cross-linked using hydrochloric acid (HCl) vapor/freezing dry approach or ethanol/HCl/water mixture immersion for the fabrication of water-resistant HA nanofibers [66]. The unique features of the electroblowing process for fabrication of nanofibrous membranes warrant further investigation, especially for more advanced operations such as pattern formation and layer-by-layer processing.

In electroblowing, a very larger volume of air and solvent mixture is generated. This will require the solvent retrieval step, if the solvent is not water. Without such a precaution, the electroblowing process will not be environmentally friendly. For this purpose, a solvent trapping system, suitable to recover most routine solvents, based on the concept of distillation for solvent recycling can be installed in the electroblowing process.

2.2.4   Double-Jet Electrospinning

The mechanical strength of electrospun nanofibrous membrane is one of the major concerns for practical applications of these materials, especially when the high porosity of the membrane needs to be maintained in the filtration process [49, 67, 68]. To address this concern, a double-jet electrospinning process has been developed and a schematic configuration of the setup for producing composite nanofibers is illustrated in Fig. 2.5 [69].

In a double-jet electrospinning operation, two different polymer solutions are spun onto the same collector (Fig. 2.5a), where the resulting membrane contains one component that can be considered as the skeleton, and the other component forms a finer 3-D fibrous network within the skeleton scaffold, as shown in Fig. 2.5b. The thicker skeleton nanofibers provide mechanical properties for the membrane, and the thinner nanofibers offer functionality. As a result, the mechanical properties

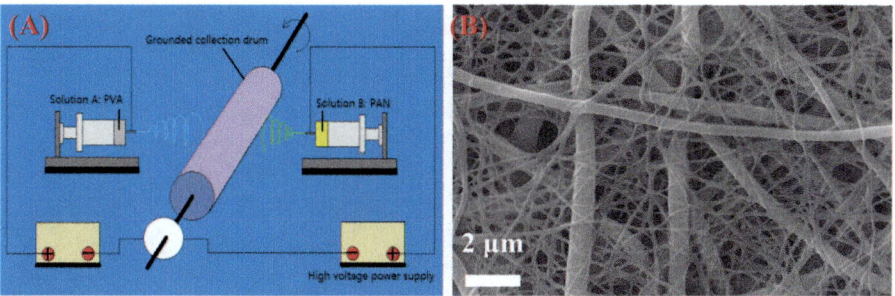

Fig. 2.5 Configuration of the double-jet electrospinning setup (**a**) and composite nanofibers made by double-jet electrospinning (**b**)

of the membrane can be essentially enhanced and used as a self-standing membrane for water purification.

As an example, PVA-PAN nanocomposite fibrous membrane was fabricated by double-jet electrospinning, where PVA nanofibers with a mean diameter of 800 ± 90 nm and PAN nanofibers with a mean diameter of 150 ± 40 nm were achieved, respectively [69]. The bimodal fiber diameter distribution of the nanocomposite membrane is observed by SEM in Fig. 2.5b. This membrane system was post-treated by GA to cross-link the PVA component, and the PAN component was also modified with polyvinylamine (PVAm) to create functional groups for heavy metal ion adsorption [70]. The cross-linked PVA nanofibrous scaffold exhibited outstanding mechanical properties which could withstand typical microfiltration operating pressure (e.g., 0.2–2.0 psi) without changing the porosity of the membrane [71]. Meanwhile, the finer PAN nanofibrous scaffold offered a high surface-to-volume ratio which was effective to improve the adsorption capacity of the membrane.

Although the single-jet and double-jet electrospinning operations are ideal for laboratory operations, the total yield of the production is relatively low. The real commercialization of electrospin products requires the scale-up production of the demonstrated electrospinning process. Therefore, the development of multiple-jet electrospinning technology became an essential step for applications of nanofibrous membranes [5].

2.2.5 Multiple-Jet Electrospinning

Multiple-jets with designed array patterns have been used to ensure the fabrication of uniform membrane thickness, where its production rate can be increased thousands of times over that of a single jet, depending on the amount of the spinnerets used [72, 73]. The multiple-jet operation, as shown schematically in Fig. 2.6, requires specially designed spinneret heads to achieve the desired electric field configuration and to isolate unwanted electrostatic field interactions. Meanwhile,

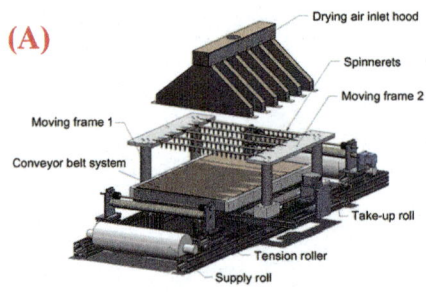 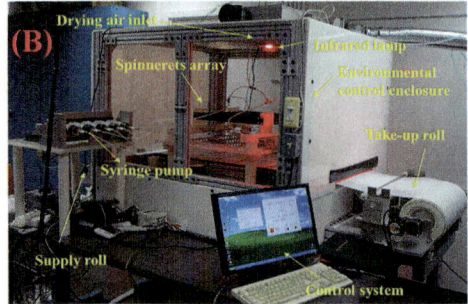

Fig. 2.6 Schematic diagram of a multiple-jet electrospinning setup (**a**) and a photograph of a lab-scaled 100-jet electrospinning apparatus (**b**). Images courtesy of Dr. Dufei Fang, Stony Brook University [74–76]

additional facilities, such as dry air inlet hood, heating element, and environmental control chamber, are essential to control the solvent evaporation rate and to ensure the quality of the large-scale production.

For example, in one multiple-jet electrospinning setup, an electrified secondary electrode, in the shape of a metal ring around each spinneret, has been used to minimize the interactions between adjacent electrodes, and to shield the primary electrodes in order to form more uniform electrical field distribution [74–76]. Using the multiple-jet electrospinning apparatus, an electrospun nanofibrous membrane can be produced continuously over a reasonably long time period. In addition, the multiple-jet electrospinning setup can also be modified to provide the flexibility to fabricate multicomponent composite nanofibrous scaffolds. For example, a combination of multiple-jet electrospinning technology and melt-blown process has been proposed to prepare micro-nanofibrous scaffolds, where the large diameter fibers from melt blowing could form a skeleton and the smaller diameter fibers from electrospinning could form the nanofibrous network. The composite membrane will be ideal for combined microfiltration and adsorption applications. In the multiple-jet electrospinning operation, the quality control of electrospun membrane fabrication has become the most important issue. The evaluation tasks typically include (1) uniformity and consistency of the produced fibers; (2) desired fiber diameter and distribution; (3) practical operational parameters; and (4) standard method development for quality monitoring.

In the multiple-jet electrospinning operation, the charged jets are repulsive to each other. Therefore, the instrument should be designed to minimize the electric field interferences between each jet. In addition to the secondary electrode approach [74–76], this challenge can be overcome by using another approach, involving the adopting of needleless electrospinning techniques, where the electrode for the formation of charged jet could be in the form of a wire, cylinder, disc, ball, or spiral coil wire, where the multiple jets were formed spontaneously on the electrodes. The typical yield of the nanofiber production by these techniques can be more than 260 times in weight compared to that of a single-jet electrospinning [72, 77].

Table 2.1 Current companies with large-scale electrospinning products [80]

USA-based companies	Foreign-based companies
Donaldson Company Inc	Toyobo (Japan)
Argonide Corporation	Toray Industries, Inc (Japan)
E-Spin Technologies Inc	Teijin Fibers, Ltd (Japan)
Nanostatics Corporation	Kato Tech Co., Ltd (Japan)
Foster Miller Inc	Espinex Inc (Japan)
KX industries	NanoNC (Korea)
EHS Technologies	Samshin Creation Co., Ltd (Korea)
SNS Nanofiber Technology, LLC	Elmarco (Czech Republic)
Hollingsworth & Vose	Nanoval GmBH & Co. (German)
Finetex Technology Inc	Nanofiber Future Tech. Corp (Canada)
Hitco Carbon Composites, Inc	Nicast Ltd (Israel)
US Global Nanospace, Inc	Ahlstrom (Finland)

Unfortunately, the nature of the spontaneous fiber formation do not guarantee the uniformity on fiber diameters and fiber distribution, whereby the resulting membranes often do not have a homogeneous pore distribution necessary for water filtration. These membranes, however, are ideal for air filtration application, where the separation mechanism is dominated by the particle/nanofiber adsorption.

In the multiple-jet electrospinning operation, a large volume of solvent will need to be recycled or disposed properly, especially for the systems involving organic solvents like DMF, N,N-dimethyl sulfoxide (DMSO), chloroform, and tetrahydrofuran (THF). Earlier, we have discussed a trapping system to recycle organic solvent, such as DMF, from the solvent/air mixtures. If the solvent recycling process is not economically effective, one practical approach is that the exhaust solvent/air mixture can be burned directly by an incinerator before charging to the atmosphere.

2.2.6 Electrospinning Technology: Current Commercial Status

Currently, there are more than 200 universities and research institutes working on varying aspects of material developments and applications involving the use of electrospinning technologies, where more than 1000 scientific papers and patents have been published every year [78, 79]. Moreover, numerous industrial companies have developed their own electrospinning technologies for large-scale nanofiber production around the world, such as USA, Japan, Korea, Germany, Czech Republic, Canada, and Finland. Some selected names of these companies are illustrated in Table 2.1.

These companies all developed their own electrospinning facilities and could produce large-scale nanofibrous products based on their own market needs for specific applications. The flexibility of the electrospinning technology further facilitated some very innovative research and development of new materials and emerging applications by researchers in universities and research institutions. Typical applications of electrospun nanofibers in the industry include air and water filtration, composite materials, aerospace industry, healthcare, energy, and cosmetics, although the applications of air and gas filtration remain to be the largest segment [79, 81–84]. For example, Donaldson has commercial branches in more than 30 countries, where their electrospun nanofiber enhanced filters have been used quite extensively for gas/liquid separation and air filtration (e.g., filters for dust collection, gas turbine/heavy-duty engine air filtration) [85]. In another example, Elmarco has offered unique needless electrospinning technology that allows the production of nanofiber textiles on an industrial scale with high production rate. They have demonstrated electrospun products with sound absorption characteristics and for the elimination of mechanical/biological impurities in polluted air [86]. As electrospun membranes can offer some superior properties (e.g., higher porosity, large surface-to-volume ratio, and more functionality) over commercial membranes, we expect the large-scale commercialization of electrospun membranes for water purification is right around the corner. We believe the application of electrospun membranes for varying water treatments will likely be the next highlight in the membrane industry.

2.3   Unique Properties of Electrospun Nanofibrous Membranes

2.3.1   Tunable Structural Characters

The major structural parameters of electrospun nanofiber membranes, pertinent to water treatment applications, are surface area, porosity, and pore size. Typically, electrospun membranes have very high surface area, high porosity, and adjustable pore size and pore size distribution that can meet the requirements for water treatments [11]. These parameters will be discussed next as how they can be controlled by the electrospinning technology.

First of all, the surface-to-volume ratio of electrospun nanofibers is usually higher than conventional membranes made by the phase-inversion or solution casting techniques [24]. For example, the typical fiber diameter of an electrospun nanofiber is about a few hundred nanometers, while the range can span from a few tens of nanometers to microns. Correspondingly, the surface area of electrospun nanofibrous membrane can be from 10 to 40 m^2/g [77]. As the surface-to-volume ratio is inversely proportional to the radius of the fiber, the thinner the fiber diameter, the higher the surface area. The large surface-to-volume ratio of the nanofibrous scaffolds is very beneficial to the adsorption of toxins, such as viruses, dyes, and heavy metal ions, to the fiber surface if properly modified. The duel functionality of filtration and adsorption allow the electrospun membranes to carve up a unique niche in the varying filtration applications [87]. The diameter of the electrospun nanofiber

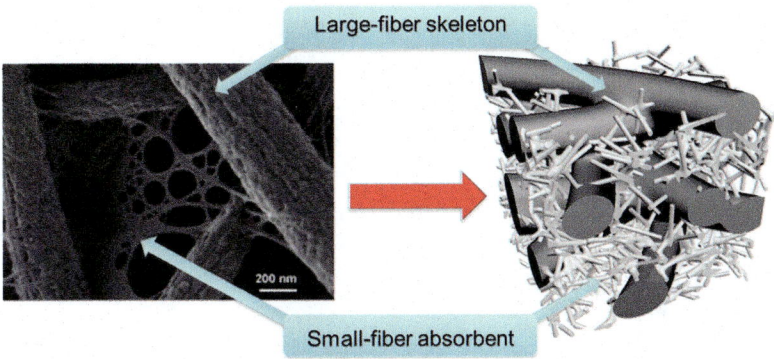

Fig. 2.7 Representation of the nanostructure of the electrospun PAN scaffold infused with very fine cellulose nanofibers

can be controlled by changing the electrospinning conditions, which means that the surface-to-volume ratio can be modulated by changing the fiber diameter. Perhaps, the easiest way to adjust the fiber diameter is by adjusting the polymer concentration during electrospinning. Typically, thinner fiber diameters can be obtained by using lower solution concentration, providing that the viscosity of the polymer solution is sufficiently high to be spinnable. The fiber diameter can also be changed by using other parameters such as higher electrical filed, larger ionic strength, and longer spinneret-to-collector distance.

For adsorption applications, one critical issue is the density and distribution of functional groups on the surface of the carrier scaffold [23]. Usually, the electrospun nanofiber scaffold having smaller fiber diameter possesses higher surface area, which can lead to higher adsorption capacity, if fiber surface is properly modified with effective adsorption sites. The smaller fiber diameter can also result in a smaller average pore size and pore size distribution of the membrane; however, this may often weaken the membrane strength. One way to overcome this problem is the use of a composite membrane structure, containing two interpenetrated fibrous networks of different fiber diameters. In one example study, very fine cellulose/chitin nanofibers (diameter about 5 nm) were infused into electrospun nanofibrous scaffolds (diameter about 200 nm), where the demonstrated composite membrane was found to have a flux two times higher than the commercial GS0.22 microfiltration membrane, and it also exhibited the adsorption capacity (against crystal violet, a cationic dye) 16 times higher than GS0.22 [44, 88, 89]. Figure 2.7 illustrates the SEM image and the schematic diagram of this composite nanofibrous membrane structure, where the very fine cellulose nanofibers significantly enhance the surface-to-volume ratio by forming a 3-D nanoweb network in the electrospun nanofiber scaffold [45]. It was found that the adsorption isotherms of this composite nanofibrous membrane followed the Langmuir model, which means that the adsorption of dye molecules was probably at the coverage of a monolayer. Meanwhile, the adsorption was found to approach the equilibrium very quickly due to the high porous structure of the membrane.

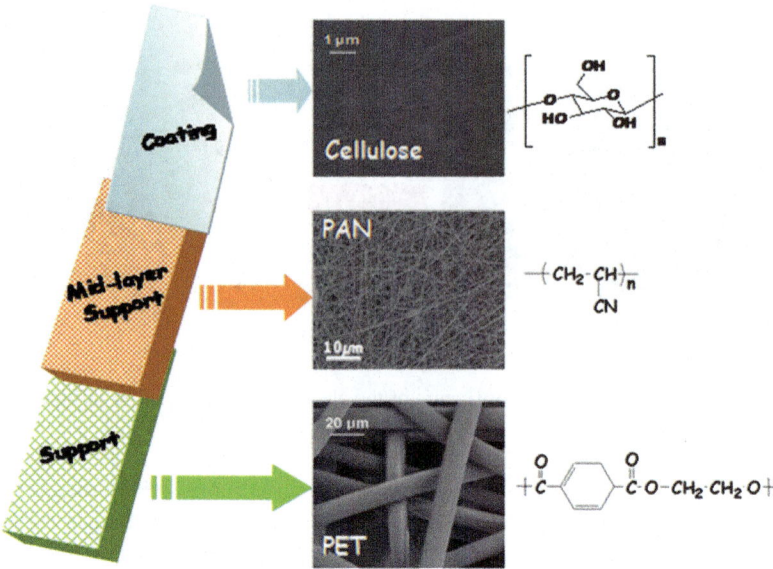

Fig. 2.8 Hierarchical structure of cellulose-based TFNC ultrafiltration membrane

The high porosity of electrospun nanofibrous membrane is beneficial for improving the permeation flux, as water molecules can transport through the membrane with low hydraulic resistance. This membrane system has an additional advantage, that is, all pores, defined by randomly deposited nanofibers, are interconnected with channel structure [25], which can reduce the hydraulic resistance of water and decrease the fouling tendency. Typically, the porosity of electrospun nanofibrous scaffold is 70–95%, depending on the electrospinning conditions [41]. On the other hand, conventional membranes fabricated by the phase-inversion approach often possess the bulk porosity from 50% to 70% with very low surface porosity 20–40% due to the asymmetric structure. The nature of high porosity makes the electrospun nanofibrous membrane as a high flux supporting scaffold to host a barrier layer for ultrafiltration (UF), nanofiltration (NF), and reverse osmosis (RO) and forward osmosis (FO) membrane fabrication [11]. For the FO operations, it was interesting to note that the concentration polarity near the barrier structure could be reduced due to the free diffusion of the salt ions in the nanofibrous scaffold.

For the UF application, a three-layered TFNC membrane, containing a PET non-woven mat as the bottom layer, electrospun PAN nanofiber scaffold as the middle support, and cellulose regenerated from ionic liquid solution as the barrier, is used, as shown in Fig. 2.8 [49]. This TFNC membrane exhibited a ten times higher permeation flux against oily bilge water than that of commercial PAN10 UF membrane, while the rejection ratio remained about the same, after 100-h filtration performance [51]. One major difference between the TFNC membrane and PAN10 was the supporting layer, where the TFNC membrane used the electrospun nanofibrous scaffold as the mid-layer support and the PAN10 membrane contained the porous

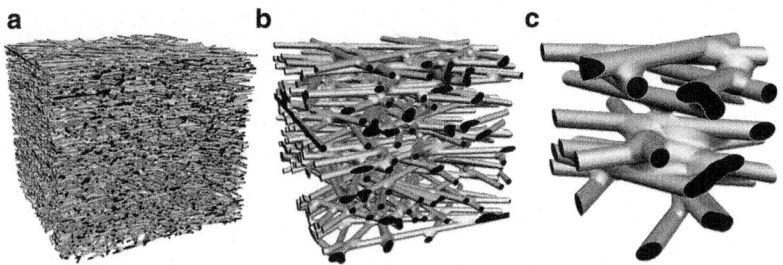

Fig. 2.9 Illustration of the correlations between pore size and fiber diameter at a constant porosity of 80 vol% in a fixed volume. The relative fiber diameter ratio of (**a**), (**b**), and (**c**) is 1:3:10

layer fabricated by the phase-inversion method. The high porosity of the electrospun nanofibrous scaffold in the TFNC membrane clearly enables the high flux membrane performance in UF, where the major separation mechanism is through size exclusion. It was also interesting to find that, the fouling trend of the TFNC membrane was relatively lower than that of PAN10, mainly due to the hydrophilic nature of its cellulose barrier layer [49].

In addition to the advantage of higher porosity, the pore size and pore size distribution of the electrospun nanofibrous membrane can be controlled by the electrospinning conditions. The pore size and pore size distribution are other two essential parameters associated with the filtration efficiency [41, 44]. From our experimental results, for typical nonwoven electrospun nanofiber scaffolds, when the fibers are randomly deposited without preferred orientation, and the porosity of the fibrous mat is kept constant at around 80 vol%, the pore size can be correlated with the fiber diameter for the membranes having optimum filtration efficiency (i.e., highest flux and highest rejection ratio). The relationship between the mean membrane pore size and the mean fiber diameter is illustrated in Fig. 2.9 [23]. In this figure, it was found that he mean pore size was about 3 ± 1 times the mean fiber diameter, and the maximum pore size was about 10 ± 2 times the mean fiber diameter. This figure can be used as a blueprint to design the structure of nanofibrous membrane, where the pore sizes can be adjusted by controlling the fiber diameter of the electrospun nanofibers.

For microfiltration, the electrospun nanofibrous scaffolds have been designed with optimal pore size and pore size distribution as a barrier [90]. For example, to purify bacteria-contaminated water, the mean pore size of the membrane has been designed to be around 0.2 μm which can remove typical waterborne bacteria such as *E. coli* and *B. diminuta*, all having the dimensions larger than 0.2 μm [91]. Based on the empirical relationship between the mean membrane pore size and the mean fiber diameter, an electrospun nanofibrous PAN membrane having the fiber diameter of around 0.1 μm but with a slightly large membrane thickness has been fabricated. The resulting membrane exhibited 6 LRV against *E. coli*, which met the criteria for drinking water purification, yet still possessing a flux about two times higher than the conventional microfilter [41, 42]. In another study [71], a suspension containing polystyrene nanoparticles with diameter of about 0.2 μm (considered as a model of bacteria-contaminated water) was used to challenge the filtration performance of an

Fig. 2.10 Electrospun
nanofibrous microfiltration
membrane after filtration
of nanoparticle suspension.
Reprinted from reference
71, Copyright (2013), with
permission from Elsevier

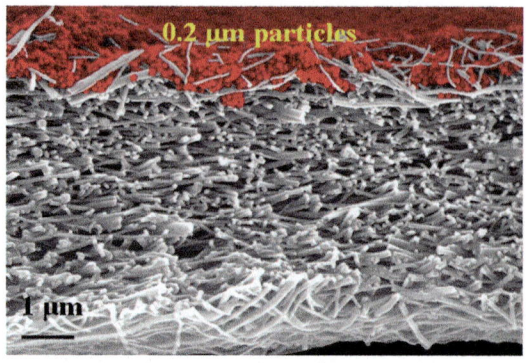

electrospun cross-linked PVA nanofibrous membrane. It was confirmed that all
nanoparticles could be rejected and aggregated on the surface of the membrane after
filtration without leaching out any to the bottom of the membrane (as seen in the
SEM image in Fig. 2.10). This indicates that the electrospun nanofibrous membrane
with suitable pore size and pore size distribution is ideal microfiltration media for
removal of bacteria particles or contaminants of similar dimensions [71, 90, 92–94].

Another application of microfiltration membranes is the MD process [95]. MD is
a thermally driven non-isothermal separation process, in which only vapor molecules can transfer through a highly hydrophobic microporous membrane. The driving force in the MD process is the vapor pressure difference induced by the
temperature difference between the feed and permeate flows. For MD operation, the
membrane surface should be hydrophobic, where the pore size of the membrane
needs to be carefully controlled to withstand a high liquid entry pressure and maintain a high (vapor) mass transfer rate. Typically, the pore size is directly proportional to the mass transfer rate, but inversely proportional to the liquid entry pressure.
In a recent study, electrospun nanofibrous polystyrene membranes have been
designed and applied for desalination of seawater and brackish water using MD
[96]. The results indicated that higher permeation flux has been achieved by using
electrospun PS membrane when compared with that of commercially available
PVDF membranes, indicating the commercial potential of using electrospun membranes for desalination applications.

Generally, as a supporting layer, the electrospun nanofibrous scaffold with
appropriate pore size can be very useful to improve the integration between the barrier layer and nanofibrous scaffold because of the high porosity of the latter, and
therefore significantly enhancing the mechanical properties of the barrier layer
under high pressure operations, such as nanofiltration and reverse osmosis [58, 97].
This is because part of the nanofibrous scaffold can be immersed in the barrier layer,
forming a nanocomposite. For example, a TFNC membrane system, involving the
use of electrospun nanofibrous PAN scaffold as the supporting layer, exhibited
excellent performance in a prolonged RO operations at 800 psi for desalination seawater [98].

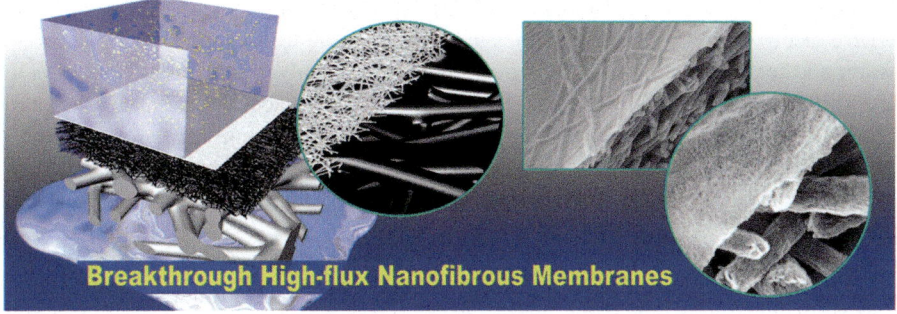

Fig. 2.11 Thin-film nanofibrous composite (TFCN) membrane containing highly porous nanofi-brous scaffold as a mid-layer [49]. Reprinted with permission from reference [49], Copyright (2011) American Chemical Society

2.3.2 Materials Selection for Nanofibrous Membranes in Water Purification

There are great deals of polymeric materials that can be electrospun to produce nanofibrous membranes for a variety of applications. However, only a few families of polymeric materials can really be used for water purification treatments [11]. Generally, these materials can be divided into three categories: hydrophilic, hydrophobic, and post-surface functionalized materials.

Typical hydrophilic polymers that are electrospun into nanofibrous membranes for water purification include PAN [41], cross-linked PVA [71], and cellulose acetate [99]. Among these materials, PAN is the most popular [100], where electrospun PAN nanofibrous membranes have been employed either as microfiltration media to screen out nanoparticles/bacteria or as the supporting scaffold to construct ultrafiltration, nanofiltration, and reverse/forward osmosis membrane [101]. The hydrophilic nature of the materials can offer tremendous advantages to the resulting membranes over conventional counterparts, antifouling property. Recently, several new types of TFNC membranes (Fig. 2.11) for microfiltration and ultrafiltration have been developed by our laboratory at Stony Brook University [43, 49, 102] and they are also being commercialized by spinoffs. These membranes are all based on the use of electrospun PAN nanofibrous scaffold as the supporting layer, where the advantages have been discussed previously.

Hydrophobic materials that are often electrospun into nanofibrous membranes for water purification include PVDF [103–105], PS [63, 94], and PES [106, 107]. These nanofibrous membranes can be used for conventional microfiltration applications, and they are also suitable to be used for MD. The conventional hydrophobic porous membranes for the MD operations include polyethylene (PE), polypropylene (PP), and polytetrafluoroethylene (PTFE) [62], where the water contact angles in these membranes are usually higher than 120°. However, PE, PP, and PTFE are difficultly to be electrospun due to great chemical resistance and very low solubility in common solvents. The only material that can be electrospun is PVDF, where electrospun PVDF membranes are very suitable for desalination of brackish water and seawater using MD. However, the available solvents to electrospun PVDF are still limited (e.g., N,N-dimethylacetamide (DMAc)/acetone or DMF/acetone) [103–105]. Recently, the successful use of electrospun PS membranes for MD operations to purify brackish water and seawater, perhaps offered a more practical commercialization pathway, as PS can be dissolved in comment solvents, such as DMF, THF, and toluene [63].

The surface properties of electrospun nanofibrous membranes only partially depend on the material properties. The membrane surface functionality can be greatly expanded in the chosen family of electrospun nanofibrous nanofibers for water purification [11, 27, 108]. The modification can be accompanied either by physical coating or chemical grafting pathways to introduce a variety of functional groups and meet the specific requirements. For example, let us consider the modification of electrospun PAN MF membranes that can remove both bacteria and viruses (the typical size is in tens of nanometers). Viruses are typically negatively charged at neutral pH value due to the isoelectrical point less than 7 (using bacteriophage *MS2* as a virus model, its isoelectrical point is only 3.9). Earlier, we demonstrated that bacteria can be removed by size exclusion through the adjustment of the membrane pore size. To remove virus (bacteriophage *MS2*), the PAN surface was modified into positively charged site that can adsorb the negative charged virus at the neutral condition. To accomplish this, PAN nanofibers could be surface-coated with a polyethylenimine (PEI)-diepoxy (EGdGE) copolymer, which provided positively charged amino-groups. The results indicate that this membrane system could adsorb *MS2* completely with 4 LRV at the pH value between 6.5 and 8.5 [43]. In another example, a polyvinyl amine hydrochloride (PVAm) having positively charged groups was grafted onto the hydrolyzed PAN surface (with carboxyl groups) amidation reaction. The amino-groups grafted nanofibrous membrane, not only could remove virus particles, but also could remove negatively charged heavy metal ions, such as chromium ions, at different pH values by chelation or charge interaction [85]. The modified membranes (either by physical or chemical means) have exhibited very high adsorption capacity, comparable or better than any current commercial products. Furthermore, these membranes can be recycled and reused many times for practical applications. We believe that almost all electrospun membranes can be modified to possess additional functionality, such as chemical adsorption ability, thus expending the functions of nanofibrous membranes as effective filtration media.

2.4 Summary and Outlook

The electrospinning technology offers opportunities to create new nanomaterials for a variety of water treatments, such as microfiltration, ultrafiltration, nanofiltration, reverse osmosis, forward osmosis, and membrane distillation. Many innovations on electrospinning instrumentation and processes have been achieved in the past two decades, from electroblowing, multiple-jet electrospinning to nanofibers functionalization, and many more will be further developed for practical applications. The flexibility in material selection and structure manipulation by electrospinning has allowed us to design and construct nanostructured membranes with characteristics that can meet requirements to deal with the specific water challenge. In this chapter, the chosen examples are meant to demonstrate the current approaches to adjust the surface-to-volume ratio, membrane porosity, pore size and distribution, materials selection, and surface functionalization in order to fabricate a wide range of purification media that can generate drinking water, and treat municipal wastewater, bilge water, produced water, brackish water, and seawater.

Acknowledgements B.S.H. thanks the financial support by the SusChEM program of the National Science Foundation (DMR-1409507) and the Electric Power Research Institute. H.M. thanks the financial support by the National Natural Science Foundation of China (51673011), the State Key Laboratory of Organic-Inorganic Composites at Beijing University of Chemical Technology (oic-201503004) and the Fundamental Research Funds for the Central Universities (buctrc201501).

References

1. Ramakrishna S (2005) An introduction to electrospinning and nanofibers. World Scientific, Singapore, pp 1–382
2. Huang ZM, Zhang YZ, Kotaki M, Ramakrishna S (2003) Compos Sci Technol 63:2223–2253
3. Lendlein A, Sisson A (2011) Handbook of biodegradable polymers: isolation, synthesis, characterization and applications. Wiley-VCH Verlag, Weinheim, pp 1–426
4. Thavasi V, Singh G, Ramakrishna S (2008) Energy Environ Sci 1:205–221
5. Fang D, Chang C, Hsiao BS, Chu B (2006) Development of multiple-jet electrospinning technology. In: Reneker DH, Fong H (eds) ACS symposium series, no. 918, Polymeric nanofibers, Chap. 7, pp 91–103
6. Burger C, Hsiao BS, Chu B (2006) Annu Rev Mater Res 36:333–368
7. http://www.polynanotec.com/news.html. Accepted 22 Sep 2011
8. Chu B, Hsiao BS, Yoon K (2008) AATCC Rev 8:31–33
9. Jayaraman K, Kotaki M, Zhang YZ, Mo XM, Ramakrishna S (2004) J Nanosci Nanotechnol 4:52–65
10. Chase GG, Varabhas JS, Reneker DH (2011) J Eng Fiber Fabr 6:32–38
11. Ma HY, Chu B, Hsiao BS (2012) In: Wei Q (ed) (Chap. 15) Functional nanofibers and applications. Wood Publishing, London, pp 331–370
12. Sill TJ, von Recum HA (2008) Biomaterials 29:1989–2006
13. Khil MS, Cha DI, Kim HY, Kim IS, Bhattarai N (2003) J Biomed Mater Res B Appl Biomater 67:675–679
14. Kumbar SG, Nair LS, Bhattacharyya S, Laurencin CT (2006) J Nanosci Nanotechnol 6:2591–2607
15. Rieger KA, Birch NP, Schiffman JD (2013) J Mater Chem B 1:4531–4541
16. Wang Z, Wan L, Liu Z, Huang X, Xu Z (2009) J Mol Catal B Enzym 56:189–195
17. Ma ZW, Kotaki M, Inai R, Ramakrishna S (2005) Tissue Eng 11:101–109
18. Prabhakaran MP, Venugopal J, Chan CK, Ramakrishna S (2008) Nanotechnology 19:455102

19. Yang X, Wang H (2010) Electrospun functional nanofibrous scaffolds for tissue engineering. In: Eberli D (ed) Tissue engineering. InTech, Rijeka, pp 159–177
20. Jo SM (2012) In: Liu J (ed) Hydrogen storage. InTech, Shanghai, pp 181–210. Chap. 8
21. Miao J, Miyauchi M, Simmons TJ, Dordick JS, Linhardt RJ (2010) J Nanosci Nanotechnol 10:5507–5519
22. Ding B, Wang M, Yu J, Sun G (2009) Sensors 9:1609–1624
23. Ma HY, Burger C, Hsiao BS, Chu B (2011) J Mater Chem 21:7507–7510
24. Ma HY, Hsiao BS, Chu B (2013) Curr Org Chem 17:1361–1370
25. Chu B, Hsiao BS (2009) J Polym Sci B Polym Phys 47:2431–2435
26. Yoon K, Hsiao BS, Chu B (2008) J Mater Chem 18:5326–5334
27. Agarwal S, Wendorff JH, Greiner A (2010) Macromol Rapid Commun 31:1317–1331
28. Strutt JW (Lord Rayleigh) (1879) Proc R Soc Lond 28:404–409
29. Zeleny J (1914) Phys Rev 3:69–91
30. Cooley JF (1902) Apparatus for electrically dispersing fluids. U.S. Patent 692,631
31. Morton WJ (1902) Method of dispersing fluids. U.S. Patent 705,691
32. Formhals A (1934) Process and apparatus for preparing artificial threads. U.S. Patent 1,975,504
33. Norton CL (1936) Method and apparatus for producing fibrous or filamentary material. U.S. Patent 2,048,651
34. Taylor G (1969) Proc R Soc Lond A 280:383–397
35. Taylor G (1969) Proc R Soc Lond A 313:453–475
36. Kumbar SG, Nukavarapu SP, James R, Hogan MV, Laurencin CT (2008) Recent Pat Biomed Eng 1:68–78
37. Zong X, Kim K, Fang D, Ran S, Hsiao BS, Chu B (2002) Polymer 43:4403–4412
38. Liu Y, Ma HY, Hsiao BS, Chu B, Tsou AH (2016) Polymer 107:163–169
39. Liang D, Hsiao BS, Chu B (2007) Adv Drug Deliv Rev 59:1392–1412
40. Homaeigohar S, Elbahri M (2014) Materials 7:1017–1045
41. Wang X, Hsiao BS (2016) Curr Opin Chem Eng 12:62–81
42. Ma HY, Hsiao BS, Chu B (2014) J Membr Sci 452:446–452
43. Wang R, Liu Y, Li B, Hsiao BS, Chu B (2012) J Membr Sci 392–393:167–174
44. US Patent, 13/018 0917, 2013; WO Patent, 12/094 407, 2012
45. Ma HY, Burger C, Hsiao BS, Chu B (2012) Biomacromolecules 13:180–186
46. Sato A, Wang R, Ma HY, Hsiao BS, Chu B (2011) J Electron Microsc 60:201–209
47. Wang X, Chen X, Yoon K, Fang D, Hsiao BS, Chu B (2005) Environ Sci Technol 39: 7684–7691
48. Yoon K, Kim K, Wang X, Fang D, Hsiao BS, Chu B (2006) Polymer 47:2434–2441
49. Ma HY, Burger C, Hsiao BS, Chu B (2011) Biomacromolecules 12:970–976
50. Ma HY, Yoon K, Rong L, Mao Y, Mo Z, Fang D, Hollander Z, Gaiteri J, Hsiao BS, Chu B (2010) J Mater Chem 20:4692–4704
51. Yoon K, Hsiao BS, Chu B (2009) J Membr Sci 338:145–152
52. Ma HY, Hsiao BS, Chu B (2011) Polymer 52:2594–2599
53. Ma HY, Hsiao BS, Chu B (2014) J Membr Sci 454:272–282
54. Ma HY, Yoon K, Rong L, Shokralla M, Kopot A, Wang X, Fang D, Hsiao BS, Chu B (2010) Ind Eng Chem Res 49:11978–11984
55. Wang Z, Ma HY, Hsiao BS, Chu B (2014) Polymer 55:366–372
56. Yoon K, Hsiao BS, Chu B (2009) J Membr Sci 326:484–492
57. Yung L, Ma HY, Wang X, Yoon K, Wang R, Hsiao BS, Chu B (2010) J Membr Sci 365:52–58
58. Wang X, Yeh TM, Wang Z, Yang R, Wang R, Ma HY, Hsiao BS, Chu B (2014) Polymer 55: 1358–1366
59. Wang X, Ma HY, Chu B, Hsiao BS (2017) Desalination 420:91–98. Accepted
60. Bui N, Lind ML, Hoek EMV, McCutcheon JR (2011) J Membr Sci 385–386:10–19
61. Lawson KW, Lloyd DR (1997) J Membr Sci 124:1–25
62. Alkhudhiri A, Darwish N, Hilal N (2012) Desalination 287:2–18

63. Li X, Wang C, Yang Y, Wang X, Zhu M, Hsiao BS (2014) ACS Appl Mater Interfaces 6: 2423–2430
64. Um IC, Fang D, Hsiao BS, Okamoto A, Chu B (2004) Biomacromolecules 5:1428–1436
65. Chu B, Hsiao B.S, Fang D, Okamato A (2008) Crosslinking of hyaluronan solutions and nanofibrous membranes made therefrom. U.S. Patent 7,323,425
66. Wang X, Um IC, Fang D, Okamoto A, Hsiao BS, Chu B (2005) Polymer 46:4853–4867
67. Kallioinen M, Pekkarinen M, Manttari M, Nuortila-Jokinen J, Nystrom M (2007) J Membr Sci 294:93–102
68. Bohonak DM, Zydney AL (2005) J Membr Sci 254:71–79
69. Liu X, Ma HY, Hsiao BS (2018) unpublished results
70. Kidoaki S, Kwon IK, Matsuda T (2005) Biomaterials 26:37–46
71. Liu Y, Wang R, Ma HY, Hsiao BS, Chu B (2013) Polymer 54:548–556
72. Niu H, Wang X, Lin T (2011) Needleless electrospinning: developments and performances. In: Lin T (ed) Nanofibers-production, properties and functional applications. InTech, Rijeka, pp 17–36
73. Agarwal S, Greiner A, Wendorff JH (2008) Polymer 49:5603–5621
74. Chu B, Hsiao BS, Fang D, Brathwaite C (2004) Biodegradable and/or bioabsorbable fibrous articles and methods for using the articles for medical applications. U.S. Patent 6685956
75. Chu B, Hsiao BS, Fang D (2004) Apparatus and methods for electrospinning polymeric fibers and membranes. U.S. Patent 6713011
76. Chu B, Hsiao BS, Hadjiargyrou M, Fang D, Zong S, Kim K (2004) Cell delivery system comprising a fibrous matrix and cells. U.S. Patent 6790455
77. Teo WE, Ramakrishna S (2006) Nanotechnology 17:R89–R106
78. Web of Science database, keywords: "electrospinning or electrospun"
79. Persano L, Camposeo A, Tekmen C, Pisignano D (2013) Macromol Mater Eng 298:504–520
80. Kaur S, Gopal R, Ng WJ, Ramakrishna S, Matsuura T (2008) MRS Bull 33:21–26
81. Kriegel C, Arecchi A, Kit K, McClements DJ, Weiss J (2008) Crit Rev Food Sci Nutr 48:775–797
82. Petrik S (2011) Industrial production technology for nanofibers. In: Lin T (ed) Nanofibers – production, properties and functional applications. InTech, Rijeka, pp 1–16
83. Fang J, Wang X, Lin T (2011) Functional applications of electrospun nanofibers. In: Lin T (ed) Nanofibers – production, properties and functional applications. InTech, Rijeka, pp 287–326
84. Ramakrishna S, Fujihara K, Teo WE, Yong T, Ma Z, Ramaseshan R (2006) Mater Today 9: 40–50
85. http://www.donaldson.com
86. http://www.elmarco.com
87. Liu Y, Ma HY, Liu B, Hsiao BS, Chu B (2015) J Plast Film Sheeting 31:379–400
88. Yang R, Aubrecht KB, Ma HY, Grubbs RB, Hsiao BS, Chu B (2014) Polymer 55:1167–1176
89. Yang R, Su Y, Burger C, Aubrecht KB, Wang X, Ma H, Grubbs RB, Hsiao BS, Chu B (2015) Polymer 60:9–17
90. Barhate RS, Ramakrishna S (2007) J Membr Sci 296:1–8
91. http://www.britannica.com/facts/5/463522/E-coli-as-discussed-in-bacteria. Facts about E. coli: dimensions, as discussed in "bacteria: diversity of structure of bacteria" in Britannica Online Encyclopedia. Accepted 22 Sept 2013
92. Kaur S, Barhate R, Sundarrajan S, Matruura T, Ramakrishna S (2011) Desalination 279: 201–209
93. Gopal R, Kaur S, Ma Z, Chan C, Ramakrishna S, Matsuura T (2006) J Membr Sci 281:581–586
94. Barhate RS, Loong CK, Ramakrishna S (2006) J Membr Sci 283:209–218
95. Woods J, Pellegrino J, Burch J (2011) J Membr Sci 368:124–133
96. Ke H, Feldman E, Guzman P, Cole J, Wei Q, Chu B, Alkhudhiri A, Alrusheed R, Hsiao BS (2016) J Membr Sci 515:86–97
97. Wang X, Fang D, Hsiao BS, Chu B (2014) J Membr Sci 469:188–197

98. Ma HY, Hsiao BS (2018) High-flux thin-film nanocomposite reverse osmosis membrane for desalination. U.S. patent. 2018/0508903
99. Ma Z, Kotaki M, Ramakrishna S (2005) J Membr Sci 265:115–123
100. Nataraj SK, Yang KS, Aminabhavi TM (2012) Prog Polym Sci 37:487–513
101. Ahmad FE, Lalia BS, Hashaikeh R (2015) Desalination 356:15–30
102. Chu B, Hsiao BS, Ma HY (2016) High flux high efficiency nanofiber membranes and methods of production thereof. U.S. Patent 9511329
103. Dong ZQ, Ma XH, Xu ZL, You WT, Li FB (2014) Desalination 347:175–183
104. Prince JA, Singh G, Rana D, Matsuura T, Anbharasi V, Shanmugasundaram TS (2012) J Membr Sci 397:80–86
105. Essalhi M, Khayet M (2013) J Membr Sci 433:167–179
106. Yoon K, Hsiao BS, Chu B (2009) Polymer 50:2893–2899
107. Tang Z, Qiu C, McCutcheon JR, Yoon K, Ma HY, Fang D, Lee E, Kopp C, Hsiao BS, Chu B (2009) J Polym Sci B Polym Phys 47:2288–2300
108. Agarwal S, Greiner A, Wendorff JH (2013) Prog Polym Sci 38:963–991

Chapter 3
Electrospun Filters for Air Filtration: Comparison with Existing Air Filtration Technologies

Yan Wang, Xu Zhao, Xiuling Jiao, and Dairong Chen

Abstract Particulate matter (PM) in air is a severe threat to the health of humankind. The existing air filtration technologies, which mainly include the porous film filter and the conventional fibrous filter prepared by meltblown or spunbonded process, could not simultaneously possess high filtration efficiency and low pressure drop toward fine particulates. Developing novel air filters with excellent filtration performance is of great importance. Electrospun nanofibrous membranes attract much attention due to their many fascinating characteristics, such as small fiber diameter, large open porosity, and their wide raw material sources. Thanks to the abovementioned properties, electrospun filters with the electrospun nanofibrous membranes as the core filtration media are intensively studied. A large amount of raw materials including polymers and ceramics have been fabricated to nanofibrous membranes via electrospinning, and they are applied as air filters. These electrospun filters exhibit attracting characteristics such as high filtration efficiency, low pressure drop, and good capturing ability toward fine particulates. Moreover, the electrospun filters exhibit great potentials in many applications, such as industrial dust filtration, locomotive air filtration, and indoor air filtration. In this chapter, the history and filtration mechanism, the main materials, and the applications of the electrospun filters are introduced, and the comparison of the electrospun filters with the existing air filtration technologies is discussed. This chapter may shed some light on the development of the electrospun filters for eliminating PM pollutions from air.

Y. Wang · X. Zhao (✉)
Key Laboratory of Drinking Water Science and Technology, Research Center
for Eco-Environmental Sciences, Chinese Academy of Sciences, Beijing, P.R. China
e-mail: yanwang@rcees.ac.cn; zhaoxu@rcees.ac.cn

X. Jiao · D. Chen
School of Chemistry and Chemical Engineering, National Engineering Research Center
for Colloidal Materials, Shandong University, Jinan, P.R. China
e-mail: jiaoxl@sdu.edu.cn; cdr@sdu.edu.cn

© Springer International Publishing AG, part of Springer Nature 2018
M. L. Focarete et al. (eds.), *Filtering Media by Electrospinning*,
https://doi.org/10.1007/978-3-319-78163-1_3

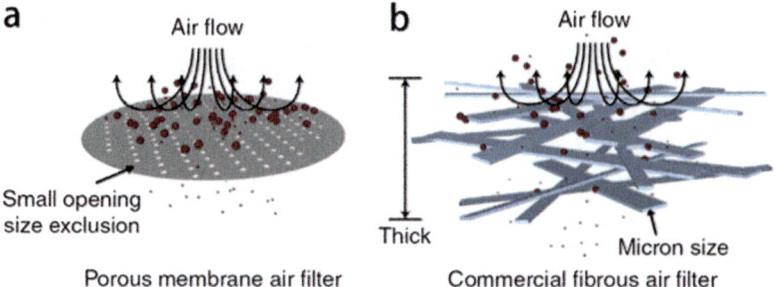

Fig. 3.1 (**a**) Schematics of porous air filter capturing PM particles by size exclusion. (**b**) Schematics of bulky fibrous air filter capturing PM particles by thick physical barrier and adhesion. (Reproduced with permissions from Ref. [10])

3.1 Introduction

Air pollution, as a result of the rapid urbanization and industrialization, has become a great threat to mankind in recent years [1, 2]. The particulate matter (PM) suspended in air is one of the main contributions to the air pollution, and is considered to be associated with the diseases such as retinal microvascular changes, atherosclerosis, lung cancer, and cardiovascular diseases [3–6]. Air filtration is becoming an urgent need for improving air quality and reducing the incidence of diseases caused by PM [7, 8].

The air filters could remove the particulate matter suspended in air, and the filtration performance of the filters is closely associated with the air filtration media. Among the existing air filtration technologies, there are two types of commonly used air filtration media, which includes the porous film filter and the fibrous filter [9]. As shown in Fig. 3.1a, the porous film filter is similar to the water filtration filter, which is made by creating pores on solid substrate. The pore size of the porous film filter is usually very small, and the porosity is low (<30%). Hence, this type of filter possesses a relatively high filtration efficiency but with a terribly large pressure drop. Figure 3.1b shows the schematic fibrous air filter, which captures PM particles via the thick physical barriers as well as the adhesion (chemical interaction between the PM and the fibers). This type of filter is made of diverse thick fibers including meltblown fibers, glass fibers, and spunbonded fibers; the fiber diameters varies from several microns to tens of microns. The porosity of this filter is >70%, which endows a relatively high permeability of the filters. However, there still remain some disadvantages of the fibrous filter, such as relatively low filtration efficiency, incapable of capturing the fine particles due to the micron-sized fiber diameter [10].

Generally, the filtration performance of the fibrous filters can be improved by reducing the fiber diameter [13–16].

Electrospinning is a facile and easy-to-scale-up method to fabricate nanofibers [11, 17–19]. Figure 3.2 presents a human hair surrounded by electrospun fibers, indicating the small diameter of the electrospun fiber. Figure 3.3 shows a SEM image of cellulose acetate nanofibers electrospun on 140 μm nylon grid support [12].

Fig. 3.2 Scanning electron microscope (SEM) image of a human hair surrounded by electrospun fibers of poly(vinyl alcohol) (PVA). (Reproduced with permissions from Ref. [11])

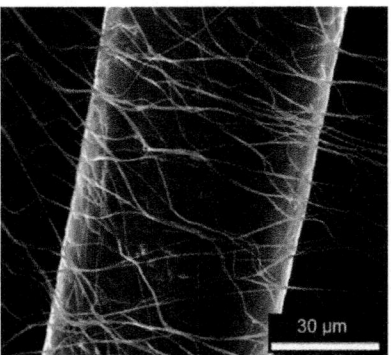

Fig. 3.3 SEM image of cellulose acetate nanofibers on 140 μm nylon grid support heated at 200 °C. (Reproduced with permissions from Ref. [12])

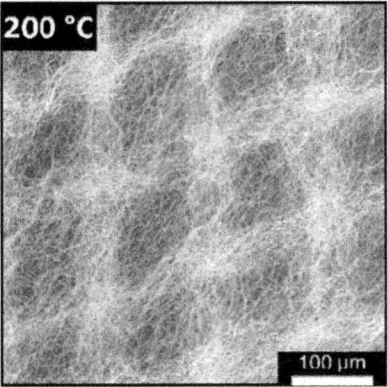

The nanofibers dominate filtration performance while the substrate provides the mechanical properties. The fiber diameters, as well as the pore sizes of the nanofibrous membrane, are much smaller than those of the substrate. The fibrous membranes produced by electrospinning method possess large surface-to-volume ratio, small fiber diameter, high porosity, and interconnected open pore structure; hence, the electrospun filters exhibit excellent filtration performance [20–24]. Herein, we present a comprehensive overview on the filtration mechanism, main materials, and applications of the electrospun filters.

3.2 The History and Filtration Mechanism of Electrospun Filters

3.2.1 The History of Electrospun Filters

The electrospinning method is intensively used for fabricating various nanofibrous filters nowadays. Fibers were successfully fabricated under electric field force since 1902 in the USA [25]. In 1934, Formhals studied the formation of the polymer

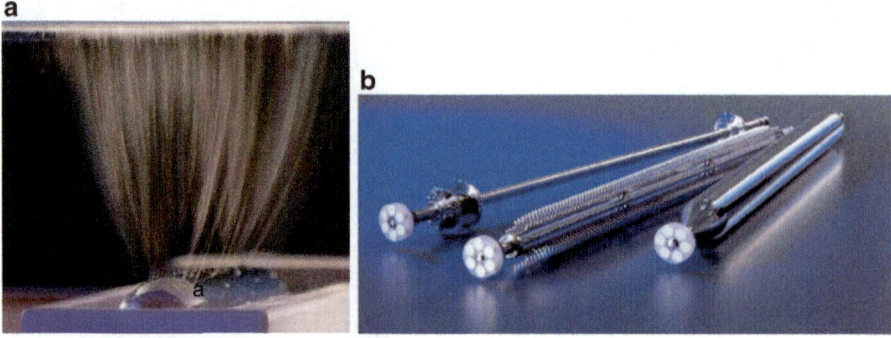

Fig. 3.4 Nozzle-less electrospinning (**a**) process with rotating electrodes, (**b**) various spinning electrodes. (Reproduced with permissions from Ref. [33])

solution jet flow between the electrodes and described the electrospinning apparatus to fabricate fibers in detail [26]. The application of electrospun fibers in filtration area was first reported by Petryanov-Sokolov in the Aerosol Department of Karpov Institute of Physical Chemistry. These materials are known as FP (filters of Petryanov) in Russia, and are now called nanofibrous filter media [27]. The scientific activities related to production of FP were announced as the top secret in Russia at that time, and the FP has been applied in nuclear energy technologies to protect the environment from nuclear active aerosol release after Second World War [28]. In Russia, there were five enterprises producing FP-related materials with an annual capacity of 20 million m^2 by the end of 1960s. In the USA, the nanofibrous materials arouse much attention and its production gained momentum in 1980 with the efforts of "Donaldson." In Europe, the electroforming fibers were produced commercially in 1990s by "Freudenberg." After years of efforts, the electrospinning method for producing fibers as air filtration materials has been applied worldwide, and more than 20 types of fiber filter materials have been produced industrially [29]. Many related properties including the effects of face velocity and nanofiber packing density on filtration performance [30], filtration modeling [31], and the mechanical properties of the membranes [32] have been studied extensively. The electrospinning instrument is also developed rapidly, which is from the single needle electrospinning, multijet electrospinning to the nozzle-less electrospinning. Figure 3.4 presents the nozzle-less electrospinning instruments for industrial production [33]. The superior filtration performance of the electrospun fibrous membranes compared to the conventional filtration media as well as the development of the electrospinning instruments contribute to the commercial production and application of the electrospun fibers.

3.2.2 Filtration Mechanism of Electrospun Filters

The study of the filtration mechanism was originated from 1827 by Robert Brown, who is a botanist in Scotland. He found the pollens and other fine particulates suspended in water were keeping irregular curve movement, which is called Brownian motion [34]. In 1931, Albrecht used the fluid mechanics to study the motion law of airflow passing a single cylindrical fiber, and established the Albrecht theory, which was then significantly improved by Shell [35]. Soon afterwards, the Brownian motion and inertia deposition were combined in the fibrous filtration mechanism, and the mathematical formula was inferred by Kaufmann [36, 37]. Meanwhile, Langmuir further studied the filtration mechanism, and considered the entrapment and diffusion were the main filtration effects [38]. The entrapment, diffusion, and inertia deposition were combined, and the isolated fiber theory was established by Davies in 1952 [39]. Thereafter, scientists continue to study the filtration mechanism intensively. Some new theories were proposed, the existing theories were optimized, and the modeling for filtration efficiency and pressure drop was established [26, 40–46].

Nowadays, the filtration mechanism has been well established, and the mechanism for filtration and separation of particulate is considered to be the comprehensive effects of inertial impaction, interception, diffusion, and electrostatic attraction [47]. Figure 3.5 shows a classic illustration of the mechanism. When the particle inertia is high enough to break away from air streamlines and impact the fiber, the inertial impaction filtration occurs. When the inertia is not high enough to break away from the streamline, but the particle comes close enough to the fiber, then adhesion forces will attach the particle to the fiber, which is called as interception. The diffusion effect is based on the random and probabilistic Brownian motion of particles with diameters smaller than 0.5 μm, which will cause the particle to move from the streamline and possibly engage the fiber surface. The electrostatic attraction also could force the particle diverted from the streamline and attracted to the fiber due to the electrostatic effect between the particle and fiber.

The main factors to evaluate the filtration performance include the filtration efficiency and pressure drop. The filtration efficiency is the percentage of contaminant removed by the filters, and the pressure drop is the differential pressure between upstream and downstream of the filter medium [47]. Based on the filtration efficiency and pressure drop, the quality factor (QF), a comprehensive parameter for comparing filtration performance of different filters, is defined as [48]:

$$QF = \left[-\ln(1-\eta) \right] / \Delta p.$$

where η and Δp represent the filtration efficiency and pressure drop, respectively. A better filter is the one that possesses a higher filtration efficiency and/or lower pressure drop, which corresponds to a higher QF value [14, 49].

The electrospun filter exhibits excellent filtration performance toward the fine particulates due to its small fiber diameter, high surface area, high porosity, interconnected

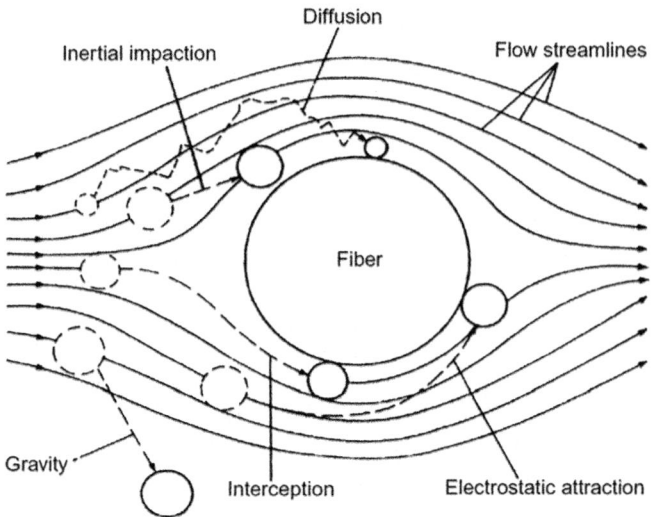

Fig. 3.5 Mechanisms of particle capture—flow past a single fiber (the flow being perpendicular to the axis of the fiber.) (Reproduced with permissions from Ref. 47)

open pore structures, and high permeability [13, 20, 24, 50]. Specifically, the electrospun filters could achieve a high filtration efficiency with a relatively low pressure drop. Compared with the existing filtration technologies, the fibers that compose the electrospun filters are with much smaller diameters (<500 nm). Under the same pressure drop, the frequencies of diffusion, interception, and inertial impaction all improved when the airflow streams around fibers with a smaller diameter. Hence, the membranes with smaller diameter fibers possess higher abilities to capture the fine particulates [51].

3.3 Materials for Electrospun Filters

3.3.1 Polymers

Linear polymers are preferred in electrospinning technique due to their appropriate rheological properties [52]. The electrospun polymer membranes are intensively studied as filtration media for vehicle cabin filters, building filters, and personal respiratory, and they can be used at temperatures lower than 150 °C for long time [53]. In the past years, a large number of fibrous membranes fabricated by electrospinning have been applied in gas filtration. These fibrous membranes involve a wide variety of materials such as Nylon-6 [14, 21], elastomeric fibers [13], polyacrylonitrile (PAN) [23, 54], polyethyleneoxide (PEO) [30], polyurethane [31], polyvinyl chloride/polyurethane [32], and Nylon-6,6 [55]. Table 3.1 presents a list of the structures of the commonly used polymers for electrospun filter materials.

Table 3.1 A list of polymers for electrospun filters

Polymer (Chemical name)	Trade name/Acronym	Structure
Polyamide (Poly(hexa-methyleneadipamide))	Nylon 6,6/PA-66	
Polyamide (Polycaprolactam)	Nylon 6/PA-6	
Polyvinylidene difluoride	PVDF	
Polyacrylonitrile	PAN	
Polyurethane	Spandex	
Polyethyleneoxide	PEO	H-(-O-CH2-CH2-)n-OH
Polylactic acid (Polylactide)	PLA	
Polyvinyl chloride	PVC	
Polyimide	P84	

Nylon is the generic name for polyamide. There are two major commercial forms of Nylon: Nylon 6,6 and Nylon 6. Nylon 6,6 is produced by the polymerization of hexamethylene diamine and adipoyle chloride, while Nylon 6 is produced by the polymerization of caprolactam, and they possess similar properties except that the melting point of Nylon 6,6 is higher than that of Nylon 6. Both Nylon 6,6 and Nylon 6 fibers have high flexibility, good resilience, easy processability, and strong tensile strength [47, 56].

Lee et al. fabricated Nylon 6 fibrous membrane with fiber diameters of 80–200 nm via electrospinning technique. As shown in Fig. 3.6, the filtration efficiency of the electrospun Nylon 6 filters is superior to the commercialized high-efficiency particulate air (HEPA) filter at the face velocity between 3 and 10 cm/s. The minimum removal efficiency of HEPA filters toward fine particles with 0.3 μm in diameter is 99.97%, and the filtration efficiency of electrospun Nylon 6 is 99.993% [21]. Ding et al. fabricated a novel PA-66 fibrous membrane consisting of a two-tier composite structure for air filtrations. As shown in Fig. 3.7, the filter is composed of traditional nonwoven polypropylene and two-dimensional (2D) PA-66 spider-web-like nanonets, which endow the filters some attracting features such as high filtration efficiency (up to 99.9%) and low pressure drop, which favor its practical applications [55].

PAN is a hydrophobic polymer and is insoluble in a large number of solvents; it is a versatile material and has been widely used as commercial membranes [57]. Figure 3.8 presents a SEM image of the electrospun PAN fibers, the diameters of the fibers are mostly around 220 nm [54]. Okuyama et al. fabricated PAN fibers with diameters in 270–400 nm range via electrospinning, and the PAN fibers were evaluated as a filter media using monodisperse NaCl nanoparticles with diameters below

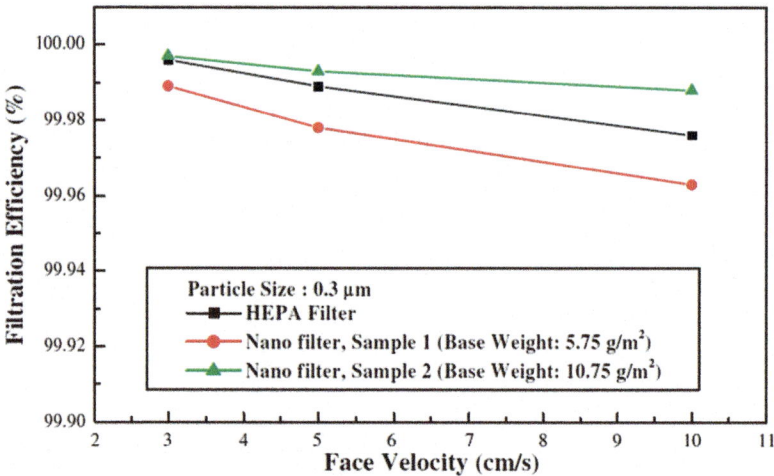

Fig. 3.6 Filtration efficiency of the HEPA filter and the Nylon 6 nanofilters as a function of face velocity. (Reproduced with permissions from Ref. [21])

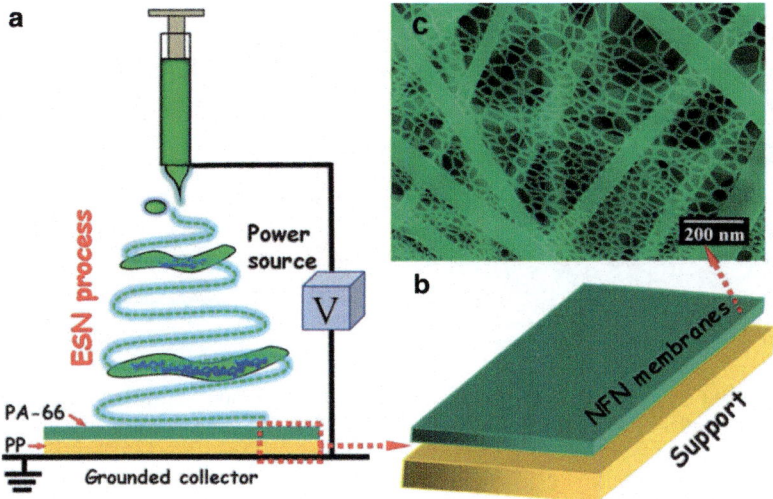

Fig. 3.7 (**a**) Schematic diagram illustrating the fabrication of PA-66 NFN membranes on nonwoven PP scaffold. (**b**) Illustration of the concept of a highly efficient filter membrane based on NFN membranes. (**c**) Typical FE-SEM image of PA-66 NFN membranes. (Reproduced with permissions from Ref. [55])

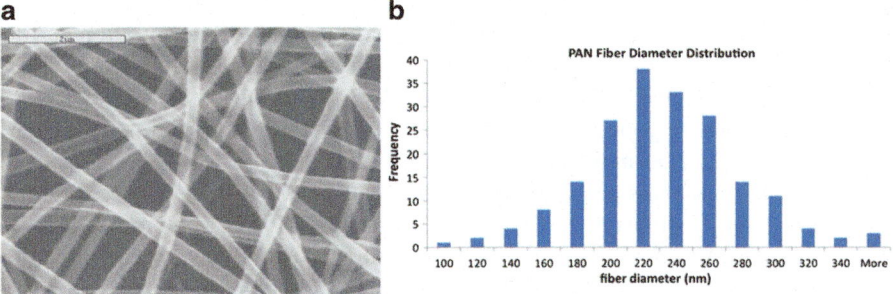

Fig. 3.8 (**a**) SEM image of PAN nanofibers and (**b**) electrospun PAN fiber diameter distribution. (Reproduced with permissions from Ref. [54])

80 nm. It was found that the electrospun PAN filters need less filter mass and possess comparable filtration efficiency compared to commercial filters [23]. Wu et al. prepared multiple thin PAN nanofiber layer mats and found that filtration performance was significantly improved compared to the HEPA filters [54].

Polyimide fibers possess unique heat resistance and good flame retardant properties. They can be used continuously at temperatures up to 260 °C. Hence, the polyimide fibers are well used for high-temperature air and fume applications. As shown in Fig. 3.9, Cui et al. developed a high-efficiency polyimide nanofiber air filter using the electrospinning technique. The electrospun polyimide filter can be used to capture PM 2.5 at high temperatures (up to 370 °C) with working time longer than 120 h.

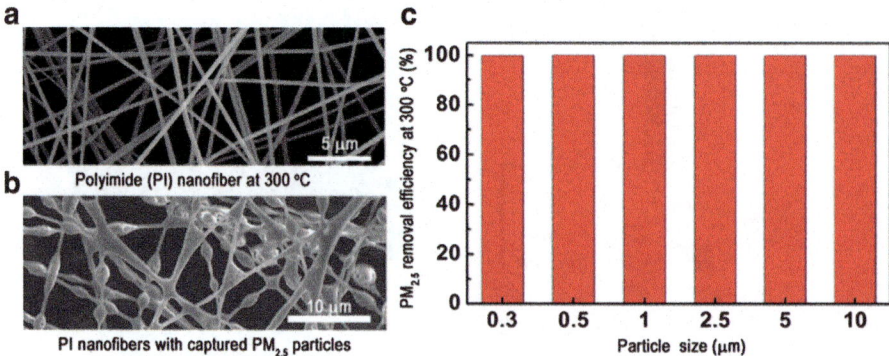

Fig. 3.9 SEM images of the electrospun polyimide nanofiber at 300 °C (**a**) with captured PM 2.5 particles (**b**), and the PM 2.5 removal efficiency at 300 °C (**c**). (Reproduced with permissions from Ref. [58])

The removal efficiency of PM particles from car exhaust at high temperature is higher than 99.5% [58].

Polylactic acid is biodegradable and it is derived from renewable resources such as cornstarch; thus, it is regarded as a sustainable material. Hence, polylactic acid has attracted much attention, and it is considered as an alternative to polyester and polypropylene. Wang et al. prepared polylactic acid/TiO$_2$ fibrous membranes via electrospinning technique, and the membranes exhibit excellent air filtration performance and good antibacterial activity. The filtration efficiency of the membrane is 99.996% with a relatively low pressure drop (128.7 Pa), and the antibacterial activity is 99.5% [59]. These good performances of the polylactic acid membranes suggest their great potentials as multifunctional filters in filtration area.

Besides the polymers described above, many other polymers such as polyvinyl-pyrrolidone (PVP), polystyrene (PS), polyvinyl alcohol (PVA), polypropylene (PP), poly(m-phenylene isophthalamide) (PMIA), and PVDF [60] have been studied as electrospinning filters. Recently, Ding et al. have fabricated a novel poly(m-phenylene isophthalamide) nanofiber/nets (PMIA NF/N) air filter via electrospinning/netting by carefully adjusting the solution composition, humidity, and additives [61]. The electrospun PMIA filter is composed of a scaffold of nanofibers and some abundant two-dimensional ultrathin (~20 nm) nanonets, and it exhibits high filtration efficiency (99.999%) and low pressure drop (92 Pa) for 300–500 nm particles. Figure 3.10 presents the performance of PM 2.5 capture by electrospun transparent air filters with various polymers. As presented in Fig. 3.10b, the PAN filters possess the best filtration performance among these polymers. The filtration efficiency of the electrospun transparent PAN filter toward PM 2.5 could maintain at 95–100% for 100 h [10]. These electrospun polymer filters exhibit excellent filtration performance and could be used to improve our living air environment.

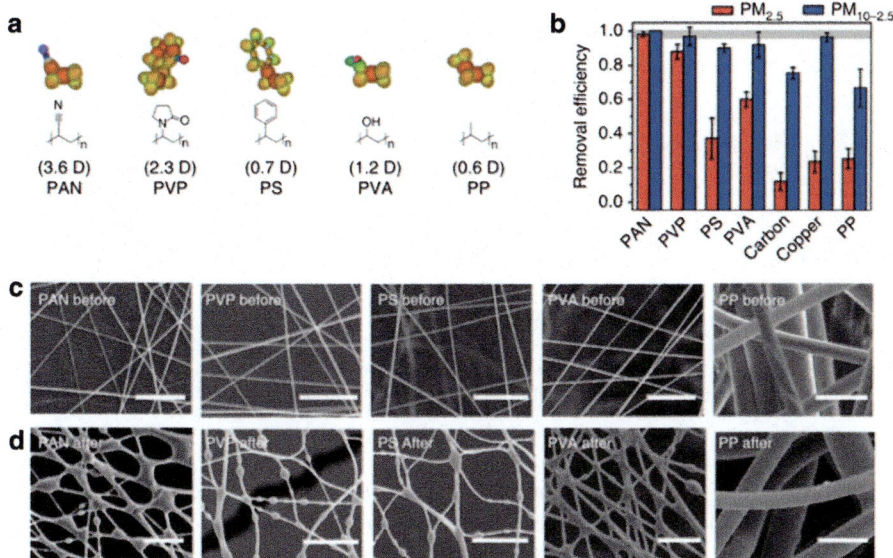

Fig. 3.10 Performance of PM 2.5 capture by transparent air filters with different surfaces. (**a**) Molecular model and formula of different polymers including PAN, PVP, PS, PVA, and PP with calculated dipole moments of the repeating units of each polymer. (**b**) Removal efficiency comparison between PAN, PVP, PS, PVA, PP carbon and copper transparent filters with same fiber diameter of ~200 nm and same transmittance of ~70%. Error bar represents the standard deviation of three replicate measurements. (**c**) SEM images of PAN, PVP, PS, PVA, and PP transparent filters before filtration. (**d**) SEM images of PAN, PVP, PS, PVA, and PP transparent filters after filtration showing the PM attachment. Scale bars in c, d: 5 μm. (Reproduced with permissions from Ref. [10])

3.3.2 Ceramics

In industrial situations, the exhaust gases from activities such as metal processing and electric power generation are at high temperatures and contain harmful particulates, which contribute substantially to air pollution. Therefore, it is worthwhile to investigate high-temperature filtration materials intensively.

Ceramic fibrous membranes possess high chemical and thermal stability and can be used under severe conditions. Figure 3.11 shows a typical procedure for electrospinning preparation of ceramic nanofibers. The preparation procedure comprises three main steps [52]: (1) preparation of the precursor, which usually contains the inorganic salt or alkoxide as source materials, and the polymers as additives to enhance the spin ability; (2) electrospinning to obtain the xerogel fibrous membranes where the precursor is usually electrospun with appropriate parameters at room temperature. The xerogel fibrous membranes composed by the inorganics and the polymer additives; and (3) transformation from xerogel fibers to ceramic fibers where the xerogel fibrous membranes are usually calcined at high temperatures to remove the polymer additives, and sometimes the amorphous inorganic membrane will transfer to crystalline ceramic membrane during calcinations.

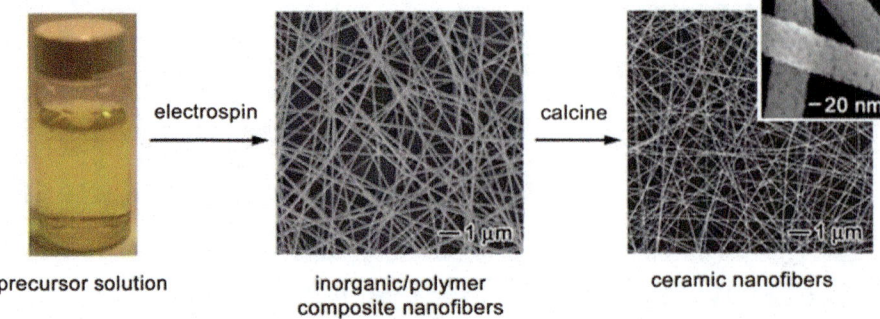

precursor solution   inorganic/polymer   ceramic nanofibers
                                       composite nanofibers

Fig. 3.11 Schematic illustration of a typical procedure for preparing ceramic nanofibers by electrospinning, the example is TiO_2 nanofibers. (Reproduced with permissions from Ref. [52])

Generally, the inorganic fibrous membranes are fragile and cannot be used as self-standing air filters. This is because during the calcination process, the solvent, organic additives, and inorganic anions in the xerogel fibers are removed, and crystallization and grain growth also occur in most cases. These processes usually lead to the generation of pores in the fiber, and consequently decrease the mechanical property of the fibers. Hence, the ceramic membranes are usually fragile.

Up until now, reports on electrospinning ceramic membranes as air filters are very limited. Only a few kinds of flexible ceramic membranes, such as SiO_2, Al_2O_3, and Pd/CeO_2–TiO_2 nanofibrous membranes have been fabricated and studied as air filters. As shown in Fig. 3.12, the self-standing γ-Al_2O_3 fibrous membranes have been successfully fabricated via electrospinning technique. The γ-Al_2O_3 membranes possess good flexibility, high tensile strength (2.98 MPa), and good thermal stability (up to 900 °C); its filtration efficiency toward 300 nm dioctyl phthalate fine particulates is 99.848% with a pressure drop of 239.12 Pa, which could meet the HEPA filter standards [62]. Chen et al. fabricated the Pd/CeO_2-TiO_2 fibrous membrane via electrospinning. The addition of CeO_2 could increase the grain boundaries and decrease the particle size in the fiber; thus, the mechanical strength is enhanced to 1.28 MPa. The Pd loaded in the fibers could endow the membranes good CO oxidation ability. The membranes could keep well at 400 °C for 20 h; the filtration efficiency is 99.86% and the pressure drop is 178 Pa [63]. SiO_2 nanofibrous membranes also have been fabricated and studied as air filters [64]. The SiO_2 membranes exhibit high flexibility (0.0156 gf cm), good tensile strength (5.5 MPa), and good thermal stability (up to 1000 °C), and excellent filtration performance toward 300–500 nm sodium chloride aerosols. These results suggest that the self-standing ceramic membranes can be obtained by choosing the appropriate precursor, adjusting the electrospinning parameters and carefully controlling the calcination procedure. The ceramic membranes exhibit great potentials for high-temperature air filtration applications.

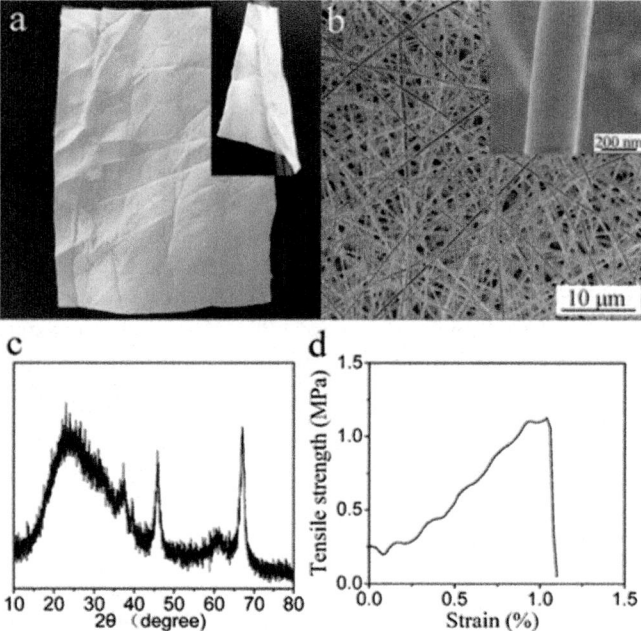

Fig. 3.12 (**a**) Optical images, (**b**) SEM images, (**c**) XRD patterns and tensile (**d**) stress–strain curves of the membranes obtained at 900 °C and further calcination at 900 °C for 24 h. (Reproduced with permissions from Ref. [62])

3.4 Applications of the Electrospun Filters

3.4.1 Industrial Dust Filtration

The industrial dust commonly arises with the industrial manufacture, such as core crushing, sieving and dosing of the powder materials, the raw materials processing of the cement and ceramics. The industrial dusts seriously affect human health, which may cause lung diseases such as lung cancer. Therefore, the air filters are needed to decrease the dust content in industrial manufacture environment to minimize its harmful effect to humankind.

For the conventional air filtration media, the industrial dusts are easy to block the filtration media, and the blocked media is hard to clean up. As a result, the pressure drop is increased seriously, and the airflow decreases to a limit value of the filtration system, hence the lifetime of the filtration media is decreased sharply. For electrospun air filters, the core filtration media is the electrospun nanofibrous membranes, which possess high dusts collecting ability, and the dusts collected on the surface of the membrane are easy to clean up. The pressure drop can recover to the initial state, and hence the lifetime of the electrospun filtration media is longer than the conventional filtration media.

The electrospun nanofibrous membranes have been successfully applied and commercialized in industrial dust filtration system now. A nanofibrous filtration material named Ultra-Web has been fabricated via electrospinning by Donaldson company in the USA [65]. The Ultra-Web was used in a special cartridge of the dust filtration system showing a filtration efficiency up to 99.999% toward 0.5 μm industrial dust. This Ultra-Web special cartridge possesses high filtration efficiency, low pressure drop, and long lifetime. Recently, a ripple-like polyamide-6 nanofiber/nets (PA-6 NF/N) air filter with extremely small pore size, highly open porous structure, and hugely extended frontal surface has been fabricated via electrospinning method [66]. This PA-6 NF/N air filter exhibits good filtration performance toward ultrafine particles, of which the removal efficiency is 99.996%, the air resistance is 95 Pa, and the quality factor is above 0.11 Pa^{-1}. The excellent filtration performance of the electrospun filters makes them to be widely used in industrial dust filtration applications.

3.4.2 Locomotive Air Filtration

The locomotive filtration is needed to guarantee the air quality for the engine and cabin. For the engine intake system, the dusts and other harmful particulates should be prevented from inhaling into the engine by the air filter. On the other hand, the air in the cabin of the locomotives such as car, plane, and train also needs purification to obtain a high quality air environment.

The conventional core filtration materials in the air filters for the engine are microporous filter paper derived from resin. This microporous filter paper is easy to be blocked, and needs maintenance and replacement frequently. Especially, the heavy weapons, such as tanker, usually work in brutal environments. The tank is easy to break down when there are much dusts in the air [67]. Electrospun nanofibrous membranes possess high filtration efficiency and low pressure drop, and hence exhibit great potential in engine intake air filtration system.

The filtration performance of electrospun Ultra-Web was studied in a Caterpillar 992 G wheel loader by Grafe et al. [65]. For the standard cellulose filter, the dust reductions of submicron particulate and the respirable (>1 micron) particulate are 68% and 86%, respectively, while for the composite cellulose/nanofiber (Ultra-Web) filter, the dust reductions of submicron particulate and the respirable (>1 micron) particulate are 92% and 93%, respectively. It can be seen that, compared to the standard cellulose filter, the filtration efficiencies of both submicron and respirable particulates are significantly increased for the Ultra-Web filter. The high filtration performance suggests the electrospun nanofibrous membranes are good candidates as engine air filters.

The electrospun nanofibrous membranes have been successfully used as engine air filters now. The air filter products based on ultrathin electrospun membranes have been developed by Donaldson company. This electrospun engine air filter possesses good filtration performance and long service life. In addition, it is suitable for the tight installation spaces and can well protect the engine. The electrospun membranes also have been successfully used in the main filter element of the M1 Abrams tank to protect its work under severe conditions. The electrospun engine air filters could improve the air quality with an ideal airflow rate, and provide a good protection for the engine.

The cabin air filter is designed to improve the air quality in the cabin of vehicles such as car, train, and airplane. For the existing cabin air filters, with the particles depositing on the filter media during filtration process, the filtration efficiency is improved, while the pressure drop increases dramatically, and hence the airflow cannot meet the requirement of the vehicles. Thus, the rapid increase of the pressure drop affects the filtration performance and lifetime of the filters. To solve the above problems, the electrospun membranes are studied to be used as cabin air filters. An electrospun polycarbonate (PC) fibrous membrane was been fabricated and used as air filter. As shown in Fig. 3.13 [68], on one hand, the filtration efficiency of the electrospun PC membrane toward 300 nm particles is above 99.98%, higher than polypropylene (PP) membrane in the existing HEPA filter; on the other hand, the pressure drops of the electrospun PC filters under various face velocity are a little higher than the PP HEPA filter, but they are still in the normal criterion range of the HEPA filter media (lower than 40 mm H_2O at 5 cm/s face velocity).

Furthermore, the electrospinning technique is easy to be combined with other techniques to prepare multifunctional cabin air filters. For example, antibacterial agents, such as Ag nanoparticles, could be dispersed in the nanofibers, and the fibers not only possess good filtration ability toward particulates, but also can prevent the invasion of bacteria. Therefore, the electrospun air filters could efficiently purify the air in cabin.

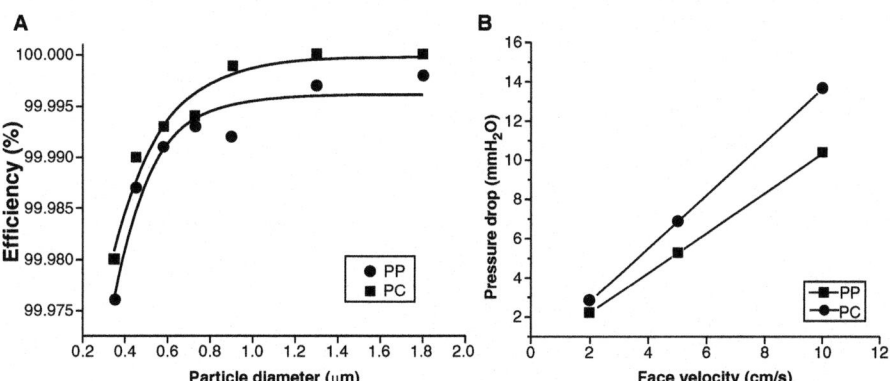

Fig. 3.13 (**a**) Filtration efficiency of PP HEPA filter and PC nanoweb according to particle diameters, (**b**) pressure drops of PP HEPA filter and PC nanoweb according to face velocities. (Reproduced with permissions from Ref. [68])

3.4.3 Indoor Air Filtration

The inhalable particles are the main particulate matter pollutants to human, which include dander, dust, spray particles, etc. In addition, some pathogenic bacteria, virus, and organic pollutants are usually adhered on the inhalable particles, which are harmful to human health. Hence, the indoor air filtration is receiving great attention. The electrospun nanofibrous membranes gain great popularity as indoor air filter media for their high filtration efficiency, large loading capacity and low pressure drop associated with the high porosity and small fiber diameter [69, 70].

The electrospun PA-6 fibrous filtration media with fiber diameters of 120–700 nm has been developed by Li et al. at Cornell University [71]. The filtration efficiency of the electrospun PA-6 membrane toward 0.5 μm particles is above 80%, higher than the conventional air filtration materials (13%). Meantime, the pressure drop maintains at a low value during the filtration process. Further study indicates that the thickness of the membranes importantly affects the filtration performance. When the package density increased from 0.03–0.50 g/m^2, the filtration efficiency increased from 42% to over 80%, and the pressure drop kept at 71.59 and 107.87 Pa. To further improve the filtration performance of the PA-6 nanofibrous membrane, Ahn et al. fabricated the PA-6 nanofibrous membrane with an average fiber diameter of 80 nm, pore diameter of 0.24 μm, and package density of 10.75 g/m^2 [21]. The filtration performance is superior to the existing HEPA filters. In addition, a dust filter bag containing electrospun nanofibrous membranes has been developed, and the dust filter bag possesses high filtration efficiency and good permeability [72]. Recently, an electrospun transparent air filter with high-efficiency PM 2.5 capturing ability has been reported by Liu et al. [10]. This transparent air filter can be installed on windows and efficiently prevent the PM from entering into the indoor environment. In all, these electrospun nanofibrous membranes could be used as indoor air filters, and will efficiently improve the indoor air quality.

3.5 Conclusions and Prospects

Electrospun filters for air filtration have been spotlighted research due to their fascinating characteristics and excellent filtration performance. Thanks to the tireless efforts of scientists on the research of electrospun filters, the basic filtration mechanism of the electrospun filters are clarified. The material types for electrospun filters are varied, from polymers to ceramics, and a large number of raw materials have been successfully tested for electrospun filters. The electrospun nanofibrous membranes have been studied as air filters in many application areas including industrial dust filtration, locomotive air filtration, and indoor air filtration. These studies provide fundamental recognition and theoretical support for electrospun filters. In this chapter, the history and filtration mechanism, the main materials and the applications of electrospun filters have been summarized. It is worth noting that the filtration performance of the electrospun filters is superior to the existing filtration technologies such as the porous film filter and the conventional fibrous filter prepared by meltblown or spunbonded processes. The outstanding performance suggests that the electrospun filters have potentially broad application in the air filtration area.

Despite the advances in the development of electrospun filters for air filtration, there are still some challenges to its further optimization and practical applications. Specifically, there are three aspects that are worth to be further investigated. Firstly, the large-scale manufacture of electrospun filters is vital for its practical applications. The studies on the electrospun air filters are mostly limited at laboratory scale. For its scale-up fabrication, industrial equipment is needed. During the large-scale production, many factors such as the concentration of the electrospinning solutions, the electrospinning parameters as well as the temperature and humidity during electrospinning process need to be adjusted accurately. Secondly, the mechanical properties of the electrospun nanofibrous membranes should be enhanced to meet the practical requirements. In applications, the high mechanical property of the membranes is preferred. The electrospun nanofibrous membranes in most cases cannot be used independently and usually need to be deposited on a nonwoven substrate during use. The development of electrospun membranes with high mechanical properties may facilitate their practical applications. Thirdly, the accurate control of the physicochemical property and microstructure (such as crystal structure) of the material is highly expected. The physicochemical properties and microstructure of the electrospun membranes significantly influence their filtration performance; thus, more theoretical and experimental results are needed even though there are still some fundamental studies up to now.

In conclusion, electrospun filters for air filtration are good candidates to remove the particulate pollutants from air, and continuous efforts are expected to push forward its intensively practical applications.

Acknowledgement This work was supported by projects of the National Key Research and Development Program of China (No. 2016YFA0203101) and the Natural Science Foundation of Beijing, China (No. 8174076).

References

1. Oberdörster G, Utell MJ (2002) Ultrafine particles in the urban air: to the respiratory tract—and beyond? Environ Health Perspect 110:A440–A441. https://doi.org/10.1289/ehp.110-a440
2. Yoon K, Hsiao BS, Chu B (2008) Functional nanofibers for environmental applications. J Mater Chem 18(44):5326–5334. https://doi.org/10.1039/B804128H
3. Barrett JR (2013) Particulate matter and cardiovascular disease: researchers turn an eye toward microvascular changes. Environ Health Perspect 121:A282–A282. https://doi.org/10.1289/ehp.121-A282
4. Dadvand P, Figueras F, Basagana X, Beelen R, Martinez D, Cirach M, Schembari A, Hoek G, Brunekreef B, Nieuwenhuijsen MJ (2013) Ambient air pollution and preeclampsia: a spa-

tiotemporal analysis. Environ Health Perspect 121:1365–1371. https://doi.org/10.1289/ehp.1206430

5. Louwies T, Panis LI, Kicinski M, De Boever P, Nawrot TS (2013) Retinal microvascular responses to short-term changes in particulate air pollution in healthy adults. Environ Health Perspect 121(9):1011–1016. https://doi.org/10.1289/ehp.1205721

6. Wilker EH, Mittleman MA, Coull BA, Gryparis A, Bots ML, Schwartz J, Sparrow D (2013) Long-term exposure to black carbon and carotid intima-media thickness: the normative aging study. Environ Health Perspect 121(9):1061–1067. https://doi.org/10.1289/ehp.1104845

7. Zhang S, Liu H, Yin X, Yu J, Ding B (2016) Anti-deformed polyacrylonitrile/polysulfone composite membrane with binary structures for effective air filtration. ACS Appl Mater Interfaces 8(12):8086–8095. https://doi.org/10.1021/acsami.6b00359

8. Eckmanns T, Rüden H, Gastmeier P (2006) The Influence of high-efficiency particulate air filtration on mortality and fungal infection among highly immunosuppressed patients: A systematic review. J Infect Dis 193(10):1408–1418. https://doi.org/10.1086/503435

9. Hinds WC (1998) Aerosol technology: properties, behavior, and measurement of airborne particles. Wiley, New York

10. Liu C, Hsu PC, Lee HW, Ye M, Zheng G, Liu N, Li W, Cui Y (2015) Transparent air filter for high-efficiency PM2.5 capture. Nat Commun 6:6205. https://doi.org/10.1038/ncomms7205

11. Greiner AW, J. H. (2007) Electrospinning: a fascinating method for the preparation of ultrathin fibers. Angew Chem Int Ed 46(30):5670–5703. https://doi.org/10.1002/anie.200604646

12. Nicosia A, Keppler T, Müller FA, Vazquez B, Ravegnani F, Monticelli P, Belosi F (2016) Cellulose acetate nanofiber electrospun on nylon substrate as novel composite matrix for efficient, heat-resistant, air filters. Chem Eng Sci 153:284–294. https://doi.org/10.1016/j.ces.2016.07.017

13. Gibson P, Schreuder-Gibson H, Rivin D (2001) Transport properties of porous membranes based on electrospun nanofibers. Colloid Surf A 187–188(0):469–481. https://doi.org/10.1016/S0927-7757(01)00616-1

14. Hung C-H, Leung WW-F (2011) Filtration of nano-aerosol using nanofiber filter under low Peclet number and transitional flow regime. Sep Purif Technol 79(1):34–42. https://doi.org/10.1016/j.seppur.2011.03.008

15. Ma H, Hsiao BS, Chu B (2011) Thin-film nanofibrous composite membranes containing cellulose or chitin barrier layers fabricated by ionic liquids. Polymer 52(12):2594–2599. https://doi.org/10.1016/j.polymer.2011.03.051

16. Li B, Cao H (2011) ZnO@graphene composite with enhanced performance for the removal of dye from water. J Mater Chem 21(10):3346–3349. https://doi.org/10.1039/C0JM03253K

17. Sridhar R, Lakshminarayanan R, Madhaiyan K, Amutha Barathi V, Lim KH, Ramakrishna S (2015) Electrosprayed nanoparticles and electrospun nanofibers based on natural materials: applications in tissue regeneration, drug delivery and pharmaceuticals. Chem Soc Rev 44(3):790–814. https://doi.org/10.1039/c4cs00226a

18. Zhang CL, Yu SH (2014) Nanoparticles meet electrospinning: recent advances and future prospects. Chem Soc Rev 43(13):4423–4448. https://doi.org/10.1039/c3cs60426h

19. Kumar PS, Sundaramurthy J, Sundarrajan S, Babu VJ, Singh G, Allakhverdiev SI, Ramakrishna S (2014) Hierarchical electrospun nanofibers for energy harvesting, production and environmental remediation. Energy Environ Sci 7(10):3192–3222. https://doi.org/10.1039/c4ee00612g

20. Gopal R, Kaur S, Ma Z, Chan C, Ramakrishna S, Matsuura T (2006) Electrospun nanofibrous filtration membrane. J Membr Sci 281(1–2):581–586. https://doi.org/10.1016/j.memsci.2006.04.026

21. Ahn YC, Park SK, Kim GT, Hwang YJ, Lee CG, Shin HS, Lee JK (2006) Development of high efficiency nanofilters made of nanofibers. Curr Appl Phys 6(6):1030–1035. https://doi.org/10.1016/j.cap.2005.07.013

22. Gopal R, Kaur S, Feng CY, Chan C, Ramakrishna S, Tabe S, Matsuura T (2007) Electrospun nanofibrous polysulfone membranes as pre-filters: Particulate removal. J Membr Sci 289(1–2):210–219. https://doi.org/10.1016/j.memsci.2006.11.056

23. Yun KM, Hogan CJ Jr, Matsubayashi Y, Kawabe M, Iskandar F, Okuyama K (2007) Nanoparticle filtration by electrospun polymer fibers. Chem Eng Sci 62(17):4751–4759. https://doi.org/10.1016/j.ces.2007.06.007
24. Ma H, Yoon K, Rong L, Mao Y, Mo Z, Fang D, Hollander Z, Gaiteri J, Hsiao BS, Chu B (2010) High-flux thin-film nanofibrous composite ultrafiltration membranes containing cellulose barrier layer. J Mater Chem 20(22):4692–4704. https://doi.org/10.1039/B922536F
25. Kanaoka C, Kishima T (1999) Observation of the process of dust accumulation on a rigid ceramic filter surface and the mechanism of cleaning dust from the filter surface. Adv Powder Technol 10(4):417–426. https://doi.org/10.1163/156855299X00253
26. Payet S, Boulaud D, Madelaine G, Renoux A (1992) Penetration and pressure drop of a HEPA filter during loading with submicron liquid particles. J Aerosol Sci 23(7):723–735. https://doi.org/10.1016/0021-8502(92)90039-X
27. Lushnikov A (1997) Igor' vasilievich petryanov-sokolov (1907–1996). J Aerosol Sci 28(4):545–546. https://doi.org/10.1016/S0021-8502(96)00473-9
28. Barhate RS, Ramakrishna S (2007) Nanofibrous filtering media: Filtration problems and solutions from tiny materials. J Membr Sci 296(1–2):1–8. https://doi.org/10.1016/j.memsci.2007.03.038
29. Shepelev AD, Rykunov VA (1995) Polymeric fiber materials for fine cleaning of gases. J Aerosol Sci 26:S919–S920. https://doi.org/10.1016/0021-8502(95)97367-N
30. Leung WW-F, Hung C-H, Yuen P-T (2010) Effect of face velocity, nanofiber packing density and thickness on filtration performance of filters with nanofibers coated on a substrate. Sep Purif Technol 71(1):30–37. https://doi.org/10.1016/j.seppur.2009.10.017
31. Sambaer W, Zatloukal M, Kimmer D (2012) 3D air filtration modeling for nanofiber based filters in the ultrafine particle size range. Chem Eng Sci 82(0):299–311. https://doi.org/10.1016/j.ces.2012.07.031
32. Wang N, Raza A, Si Y, Yu J, Sun G, Ding B (2013) Tortuously structured polyvinyl chloride/polyurethane fibrous membranes for high-efficiency fine particulate filtration. J Colloid Interface Sci 398(0):240–246. https://doi.org/10.1016/j.jcis.2013.02.019
33. Petrik S, Maly M (2009) Production nozzle-less electrospinning nanofiber technology. Mater Res Soc 1240:WW03–WW07. https://doi.org/10.1557/PROC-1240-WW03-07
34. Brown RC (1993) Theory of airflow through filters modelled as arrays of parallel fibres. Chem Eng Sci 48(20):3535–3543. https://doi.org/10.1016/0009-2509(93)85009-E
35. Davies CN (1966) Aerosol science [M]. Academic, London
36. Davies CN (1973) Air filtration [M]. Academic, London
37. McIntyre JE (2003) Synthetic fibers [M]. Woodhead, New York
38. Langmuir I (1942) Report on smokes and filters. Filtration of aerosols and the development of filter materials. Office of Scientific Research and Development No 865, SerNo 353
39. Davies CN (1979) Particle-fluid interaction. J Aerosol Sci 10(5):477–513. https://doi.org/10.1016/0021-8502(79)90006-5
40. Friedlander SK, Litt M (1958) Diffusion controlled reaction in a laminar boundary layer. Chem Eng Sci 7(4):229–234. https://doi.org/10.1016/0009-2509(58)85018-6
41. Bode HR, David CN (1979) Regulation of a multipotent stem cell, the interstitial cell of hydra. Prog Biophys Mol Biol 33:189–206. https://doi.org/10.1016/0079-6107(79)90028-2
42. Piekaar HW, Clarenburg LA (1967) Aerosol filters—the tortuosity factor in fibrous filters. Chem Eng Sci 22(12):1817–1827. https://doi.org/10.1016/0009-2509(67)80212-4
43. Clarenburg LA, Schiereck FC (1968) Aerosol filters-II theory of the pressure drop across multi-component glass fibre filters. Chem Eng Sci 23(7):773–781. https://doi.org/10.1016/0009-2509(68)85012-2
44. Brown RC, Wake D (1991) Air filtration by interception – theory and experiment. J Aerosol Sci 22(2):181–186. https://doi.org/10.1016/0021-8502(91)90026-E
45. Rosner DE, Tandon P (1995) Rational prediction of inertially induced particle deposition rates for a cylindrical target in a dust-laden stream. Chem Eng Sci 50(21):3409–3431. https://doi.org/10.1016/0009-2509(95)00161-W

46. Koeylue U, Xing Y, Rosner DE (1995) Fractal morphology analysis of combustion-generated aggregates using angular light scattering and electron microscope images. Langmuir 11(12):4848–4854. https://doi.org/10.1021/la00012a043
47. Hutten IM (2015) Handbook of nonwoven filter media, 2nd edn. Elsevier, eBook ISBN: 9780080983028, https://www.elsevier.com/books/handbook-of-nonwoven-filter-media/hutten/978-0-08-098301-1
48. Viswanathan G, Kane DB, Lipowicz PJ (2004) High efficiency fine particulate filtration using carbon nanotube coatings. Adv Mater 16(22):2045–2049. https://doi.org/10.1002/adma.200400463
49. Chen CY (1955) Filtration of aerosols by fibrous media. Chem Rev 55(3):595–623. https://doi.org/10.1021/cr50003a004
50. Yoon K, Kim K, Wang X, Fang D, Hsiao BS, Chu B (2006) High flux ultrafiltration membranes based on electrospun nanofibrous PAN scaffolds and chitosan coating. Polymer 47(7):2434–2441. https://doi.org/10.1016/j.polymer.2006.01.042
51. Kosmider K, Scott J (2002) Polymeric nanofibres exhibit an enhanced air filtration performance. Filtr Sep 39(6):20–22. https://doi.org/10.1016/S0015-1882(02)80187-2
52. Li D, McCann JT, Xia Y, Marquez M (2006) Electrospinning: a simple and versatile technique for producing ceramic nanofibers and nanotubes. J Am Ceram Soc 89(6):1861–1869. https://doi.org/10.1111/j.1551-2916.2006.00989.x
53. Wang N, Mao X, Zhang S, Yu J, Ding B (2014) Electrospun nanofibers for air filtration. In: Ding B, Yu J (eds) Electrospun nanofibers for energy and environmental applications. Springer, Berlin, pp 299–323. https://doi.org/10.1007/978-3-642-54160-5_12
54. Zhang Q, Welch J, Park H, Wu C-Y, Sigmund W, Marijnissen JCM (2010) Improvement in nanofiber filtration by multiple thin layers of nanofiber mats. J Aerosol Sci 41(2):230–236. https://doi.org/10.1016/j.jaerosci.2009.10.001
55. Wang N, Wang X, Ding B, Yu J, Sun G (2012) Tunable fabrication of three-dimensional polyamide-66 nano-fiber/nets for high efficiency fine particulate filtration. J Mater Chem 22(4):1445–1452. https://doi.org/10.1039/C1JM14299B
56. Zhang S, Shim WS, Kim J (2009) Design of ultra-fine nonwovens via electrospinning of Nylon 6: Spinning parameters and filtration efficiency. Mater Des 30(9):3659–3666. https://doi.org/10.1016/j.matdes.2009.02.017
57. Musale DA, Kumar A (2000) Solvent and pH resistance of surface crosslinked chitosan/poly(acrylonitrile) composite nanofiltration membranes. J Appl Polym Sci 77(8):1782–1793. https://doi.org/10.1002/1097-4628(20000822)77:8<1782::AID-APP15>3.0.CO;2-5
58. Zhang R, Liu C, Hsu PC, Zhang C, Liu N, Zhang J, Lee HR, Lu Y, Qiu Y, Chu S, Cui Y (2016) Nanofiber air filters with high-temperature stability for efficient PM2.5 removal from the pollution sources. Nano Lett 16(6):3642–3649. https://doi.org/10.1021/acs.nanolett.6b00771
59. Wang Z, Pan Z, Wang J, Zhao R (2016) A novel hierarchical structured poly(lactic acid)/titania fibrous membrane with excellent antibacterial activity and air filtration performance. J Nanomater 2016:1–17. https://doi.org/10.1155/2016/6272983
60. Zhao X, Li Y, Hua T, Jiang P, Yin X, Yu J, Ding B (2017) Cleanable air filter transferring moisture and effectively capturing PM2.5. Small 13 (11):1603306-n/a. https://doi.org/10.1002/smll.201603306
61. Zhang S, Liu H, Yin X, Li Z, Yu J, Ding B (2017) Tailoring mechanically robust poly(m-phenylene isophthalamide) nanofiber/nets for ultrathin high-efficiency air filter. Sci Rep 7:40550. https://doi.org/10.1038/srep40550
62. Wang Y, Li W, Xia Y, Jiao X, Chen D (2014) Electrospun flexible self-standing γ-alumina fibrous membranes and their potential as high-efficiency fine particulate filtration media. J Mater Chem A 2(36):15124. https://doi.org/10.1039/c4ta01770f
63. Li W, Wang Y, Ji B, Jiao X, Chen D (2015) Flexible Pd/CeO2–TiO2nanofibrous membrane with high efficiency ultrafine particulate filtration and improved CO catalytic oxidation performance. RSC Adv 5(72):58120–58127. https://doi.org/10.1039/c5ra09198e

64. Mao X, Si Y, Chen Y, Yang L, Zhao F, Ding B, Yu J (2012) Silica nanofibrous membranes with robust flexibility and thermal stability for high-efficiency fine particulate filtration. RSC Adv 2(32):12216. https://doi.org/10.1039/c2ra22086e
65. Timothy H, Grafe KMG (2003) Nanofiber webs from electrospinning. the nonwovens in filtration. In: Fifth International Conference, Stuttgart, Germany, March
66. Zhang S, Liu H, Zuo F, Yin X, Yu J, Ding B (2017) A controlled design of ripple-like polyamide-6 nanofiber/nets membrane for high-efficiency air filter. Small 13(10):1603151-n/a. https://doi.org/10.1002/smll.201603151
67. Timothy Grafe MG, Barris M, Schaefer J, Ric Canepa Donaldson Company Inc (2001) Nanofibers in filtration applications in transportation filtration 2001 International Conference and Exposition of the INDA (Association of the Nowovens Fabric Industry), Chicago, Illinois, December 3–5, 2001
68. Kim SJ, Nam YS, Rhee DM, Park H-S, Park WH (2007) Preparation and characterization of antimicrobial polycarbonate nanofibrous membrane. Eur Polym J 43(8):3146–3152. https://doi.org/10.1016/j.eurpolymj.2007.04.046
69. Marsano E, Francis L, Giunco F (2010) Polyamide 6 nanofibrous nonwovens via electrospinning. J Appl Polym Sci 117(3):1754–1765. https://doi.org/10.1002/app.32118
70. Supaphol P, Mit-Uppatham C, Nithitanakul M (2005) Ultrafine electrospun polyamide-6 fibers: effect of emitting electrode polarity on morphology and average fiber diameter. J Polym Sci Pol Phys 43(24):3699–3712. https://doi.org/10.1002/polb.20671
71. Lei Li MWF, Green TB (2006) Modification of air filter media with Nylon-6 nanofibers. J Eng Fibers Fabr 1(1):1–22
72. Emig D, Klimmek A, Raabe E (2002) Dust filter bag containing nano non-woven tissu. United States Patent 6395046B1

Chapter 4
Electrospun Filters for Defense and Protective Applications

Rahul Sahay

Abstract Research has experienced a rapid growth for the development of protective textiles (*PTs*) for military personnel since World War II to protect from chemical and biological warfare agents (*CBWAs*). The aim has been to fabricate PTs having full-barrier protection by degrading or blocking CBWAs. Electrospun fibrous membranes (*EFMs*) have exhibited great potential for PTs by virtue of their high surface area per unit volume, high porosity, and ability to attach functional groups for intended applications. The new generation of PTs are intended not only to adsorb but also degrade CBWAs. The aim of this chapter is to study the usage of EFMs in designing a new generation of PTs that not only provide protection from CBWAs but also provide thermal comfort to the users.

This chapter starts with the motivation for usage of EFMs in PTs. The fabrication and performance of these PTs are systematically studied by analyzing research articles focused on EFM usage in PTs. The properties of these PTs are studied with respect to (a) thermal comfort, and (b) detoxification ability against CBWAs. At the end of this chapter, a section is devoted to the progress of smart PTs. These smart PTs are envisioned to have capabilities such as sensing, self-cleaning, energy harvesting/storage, and communication.

4.1 Introduction

Typically, researchers aim to improve upon skin, an excellent natural PT, which maintains the integrity of human body against threats such as cold, heat, chemical and mechanical forces [1]. Nevertheless, the skin is highly sensitive to CBWAs such as organophosphates (*OPs*), which have been a potential hazard to the armed personnel and civilian population [2, 3]. Typically, *PTs* against CBWAs are based on full-barrier protection by adsorbing them on their surface. Nevertheless, these PTs

R. Sahay (✉)
Department of Biomaterials and Microbiological Technology, Nanotechnology Centre,
Faculty of Chemical Technology and Engineering, West Pomeranian University
of Technology Szczecin, Szczecin, Poland
e-mail: rahul.g0501308@u.nus.edu

© Springer International Publishing AG, part of Springer Nature 2018
M. L. Focarete et al. (eds.), *Filtering Media by Electrospinning*,
https://doi.org/10.1007/978-3-319-78163-1_4

are bulky and possess low thermal comfort for the wearer. Furthermore, although PTs are effective in adsorbing CBWAs, they are unable to degrade them, which can result in PT acting as a hazard in itself after decontamination. These issues motivated the researchers to develop a new generation of PTs which are lightweight, provide thermal comfort to the wearer, and are capable of selectively degrading CBWAs.

Therefore, a new generation of PTs are required to not only degrade CBWAs but also achieve esthetic and thermal comfort [1, 4–6]. Here, nanotechnology came to the rescue by the fabrication of breathable, lightweight electrospun fibrous membranes (EFMs) with desired functionalities as the new generation of PTs. In this chapter, EFM-based PTs are thoroughly analyzed. After a brief introduction, various classes of EFM-based PTs are discussed. A section is also dedicated to various smart functionalities to be incorporated in PTs, which highlight some of the challenges faced by PT industry.

4.1.1 Motivation

Since World War I, CBWAs have been widely used and stockpiled for military usage [7]. Therefore, varieties of PTs have been developed to provide effective protection against CBWAs. Although innovations in CBWAs were rapid, the development of PTs for protection from them was gradual. Protection from CBWAs started with the usage of resin oil, which was applied over the body. The resin oil being inert prevented the intrusion as well as any possible reaction with CBWAs but was not effective against aerosols. Due to widespread use of CBWAs in World War II, emphasis was placed on fabrication of PTs. Activated charcoal-based PTs were introduced, which were able to adsorb gases and aerosols and keep them trapped inside their pores. Nevertheless, permeable activated charcoal-based PTs suffer from bulkiness and can by themselves act as a hazard as PTs do not decompose CBWAs. Therefore, impermeable polymer-based PTs were developed to be resistant against aerosol and liquid unlike activated carbon-based PTs. Nevertheless, these impermeable polymers-based PTs suffer from poor moisture transmission and result in poor thermal comfort for the wearer.

The shortcoming of activated charcoal-and impermeable polymer-based PTs led to the development of varieties of lightweight, flexible, and breathable PTs. The development was spearheaded by the usage of nonwoven material in PTs. These PTs are preferred because of their exceptional absorbency, liquid repellency, resilience, and mechanical strength. Joint Service Lightweight Integrated Suit Technology (JSLIST) fabricated a two-layer PT consisting of an inner activated carbon impregnated layer covered with an outer layer of nonwoven nylon membrane. Nonwoven nylon membrane provided protection against mechanical degradation of the charcoal impregnated layer [8]. Varieties of techniques have been used to fabricate nonwoven membranes such as drawing, phase separation, template synthesis, and electrospinning. Nevertheless, electrospinning is preferred by virtue of

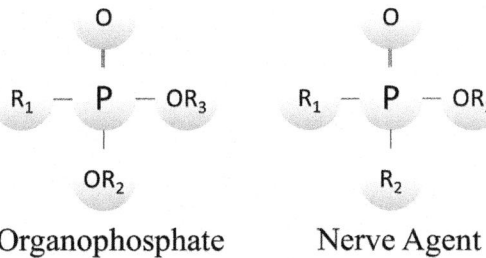

Organophosphate   Nerve Agent

Fig. 4.1 Typical structure of organophosphorus compounds: The phosphoric (P=O) bond of OPs can be substituted by 'P=S' thiophosphoric bond. 'R' represents varieties of groups that attach to basic structure. 'R$_1$' of OPs may attach via an 'O' to 'P' [14]

its versatility, control of membrane properties, and ability to be functionalized [9–11]. The EFM offers several advantages over conventional PTs not only by adsorbing but also by degrading CBWAs. Further, EFMs can be functionalized to perform two or more functions simultaneously, thus cutting down on the weight by eliminating the need for two or more separate membranes used in conventional PTs. Further, high porosity makes EFMs a possible candidate for making breathable PTs. Gibson et al. [12] noted that lightweight EFMs display appreciable breathability, airflow resistance, and improved aerosol retention in comparison to commercially available PTs. Obendorf et al. [13] also showed poly(propylene) EFM laminated textiles drastically reduce penetration of pesticides while maintaining thermal comfort. The combination of high thermal comfort and efficient barrier protection of EFMs shows their potential for the next generation of PTs.

4.1.2   Chemical and Biological Warfare Agents (CBWAs)

Exposure to CBWAs is a potential hazard to the armed personnel and civilian population. This is especially relevant to the class of CBWAs known as organophosphates (*OPs*) (see Fig. 4.1), which are produced and stockpiled for both industrial and military usage. These OPs are absorbed through inhalation or through skin to inhibit acetylcholinesterase enzyme to produce a large amount of neurotransmitter acetylcholine, which can result in paralysis or death. OPs are characterized by their volatility and persistence [15–18]. OPs can be used as gases, liquids, or solids [19–21]. Nearly ~1,300,000,000 agricultural workers are affected by OP-based pesticides worldwide. OP poisoning results in over 300,000 deaths per year worldwide [22]. Therefore, PTs are continuously being developed to effectively adsorb and detoxify OPs from contaminated surfaces.

Fig. 4.2 Methodologies
for degradation of CBWAs

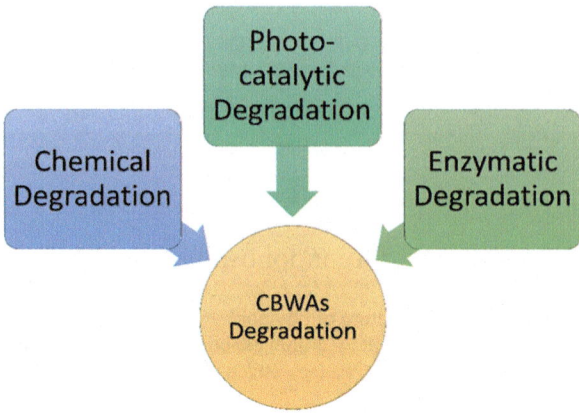

4.2 Chemical and Biological Warfare Agents (CBWAs) Degradation

Traditional PTs use breathable materials combined with activated carbon. Although these PTs offer a physical barrier against CBWAs, they cause a reduction in permeability, high heat stress, and increase in weight. Further, PTs are not able to degrade CBWAs [23]. Therefore, varieties of PTs have been designed via electrospinning to degrade these CBWAs. These PTs use methodologies such as enzymatic, photocatalytic, and chemical degradation to degrade CBWAs (see Fig. 4.2).

4.2.1 Electrospun Fibrous Membrane for CBWAs Degradation

EFMs have already been used in Ultra-Web® filter since 1981 [24]. Incorporation of EFMs in PTs can serve two main purposes: (a) protection against CBWAs, and (b) functionalities such as self-cleaning, breathability, and waterproofness [25–27]. There are mainly two methods of fabricating EFMs to offer protection against CBWAs: (a) multilayer EFM composite structure where each layer of EFM will provide a specific functionality to avoid interference between their intended operation, and (b) functionalization of EFMs to include required properties apart from degradation of CBWAs (see Figs. 4.3 and 4.4).

4.2.1.1 Chemical Degradation

Chemical degradation is the most preferred approach of all the methodologies for degradation of OPs. Alkalis can act as nucleophiles to break P-X bond in OPs resulting in nontoxic end products. Oximes can also be used as nucleophiles with high selectivity for OPs [28–30]. Further, metal oxides, such as Fe_2O_3, ZnO, TiO_2, and

Fig. 4.3 Schematic of multilayer EFM-based PT system. Each EFM will provide a specific functionality to avoid interference between their intended operation

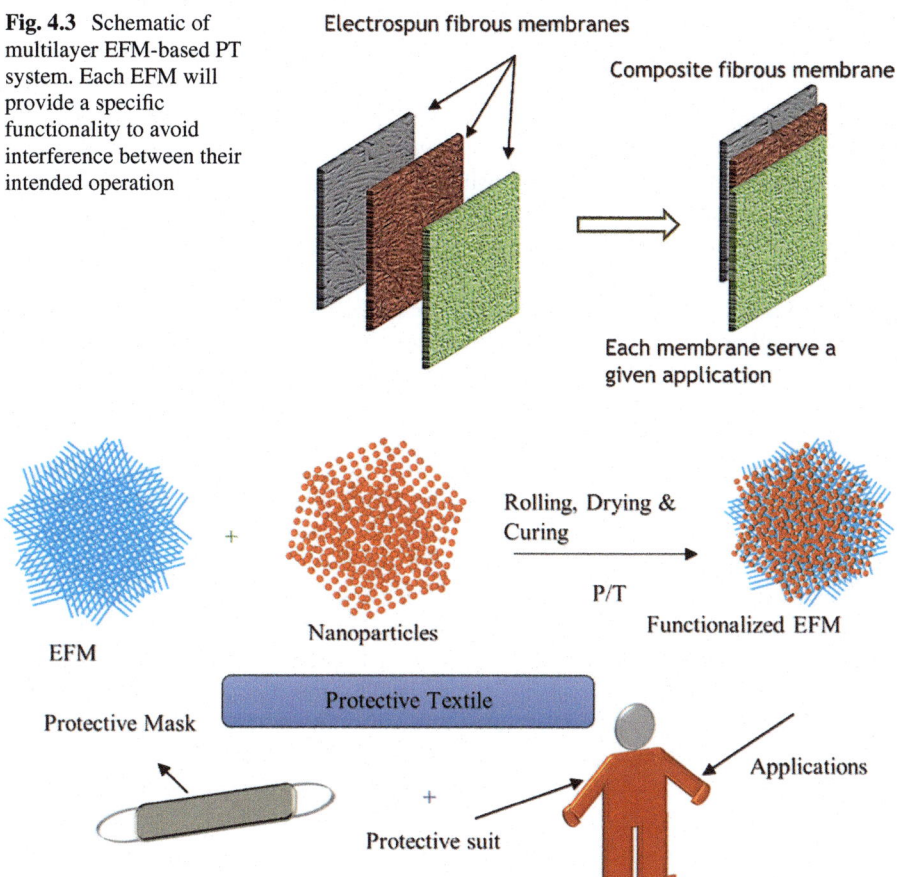

Fig. 4.4 Schematic for nanomaterial functionalized EFM. EFM can be functionalized to possess properties such as self-cleaning, breathability, and waterproofness apart from the ability to detoxify OPs

MgO, nanomaterials have shown high selectivity for hydrolysis of OPs [31–33]. These metal oxide nanomaterials were found to be more effective compared to activated carbon against OPs. Furthermore, hydrolysis of OPs is enhanced in basic or acidic environment. For example, ortho-iodobenzoic acid is widely used as a catalyst for detoxification of OPs. During hydrolysis of OPs, ortho-iodobenzoic acid is transformed into o-iodobenzoate ion, which can be easily converted back to ortho-iodobenzoic acid with oxidants such as magnesium peroxyphthalate or NaIO$_4$ [34].

The first report on EFM usage in PTs was from Donaldson Inc. and by Gibson et al. at Natick [35]. EFM was fabricated from poly(urethane), and Donaldson Inc. synthesized elastomer blended with poly(oxometalate). EFM was able to degrade 65% of 2-chloroethyl ethyl sulfide, a surrogate for mustard agent within 24 h. Chen et al. [36] fabricated multilayer poly(N-vinylguanidine)/poly(hydroxamic acid)

EFM on already fabricated poly(acrylonitrile) EFM as PT. PT was found to be effective against diisopropyl fluorophosphates, a surrogate for sarin. EFMs functionalized with chloramines, cyclodextrins, and oximes were also found to be highly effective against OPs [37]. Ramaseshan et al. [38] fabricated oxy-β-cyclodextrin from o-iodosobenzoic acid and β-cyclodextrin for incorporation into poly(vinylchloride) EFM. The composite EFM was able to hydrolyze surrogate of OP 11.5 faster than activated carbon. Ramaseshan et al. [39] also coated poly(sulfone) EFM with βCD-functionalized nanodiamonds and found it to be effective against paraoxon, a surrogate for OP. Chen et al. [40] fabricated EFM from a mixture of poly(acrylamidoxime) and poly(acrylonitrile). EFM was effective in hydrolyzing p-nitrophenyl acetate. Chen et al. [41] functionalized poly(acrylonitrile) EFM with hydroxylamine to fabricate reactive amidoxime containing EFM. The resultant EFM was able to degrade diisopropyl fluorophosphate in the presence of water. PTs based on polymeric EFMs are effective against OPs and mustard gas but are unable to provide thermal comfort to the wearer. Therefore, multilayer composite structures have also been proposed consisting of a breathable layer sandwiched between OPs protective layers [42]. Nevertheless, significant swelling could destroy the OP protective functionality of these composite EFMs.

Typically, metal oxides can detoxify OPs through oxidation and reduction reactions involving superoxide anions and hydroxyl radicals. Superoxide anions are formed by the interaction between the conduction-band free electrons and the adsorbed oxygen. Hydroxyl radicals are formed by the reaction of water with the holes in the valence band, which leads to the decomposition of OPs into nontoxic products [43, 44]. Metal oxides used in PTs include oxides of Zn, Cu, Ti, Fe, Zr, Mn, Al, V, and Ni [45–47]. Further, noble metals such as Au or Ag coated with citrate, or imidazole had been reported to degrade OPs [48, 49].

Nevertheless, these metal oxides must have high surface area to be effective against OPs. High surface area of metal oxide is achieved either by depositing metal oxide nanoparticles on EFM or by fabricating metal oxide EFMs. For instance, Ramaseshan et al. [50] fabricated $ZnTiO_3$ EFM and demonstrated its effectiveness against paraoxon, a surrogate for the OP. $ZnTiO_3$ EFM decomposed 91% of paraoxon in the first 50 min of interaction with it. Ramaseshan et al. [39] also fabricated carbon EFM impregnated with $FeTiO_3$ nanoparticles to adsorb and decompose OPs. Composite EFM exhibited excellent performance against surrogate of OP. Sundarrajan et al. [51] fabricated EFMs decorated with MgO nanoparticles for decomposition of paraoxon. EFMs employed were poly(vinylidene fluoride-co-hexafluoropropylene), poly(vinyl chloride), and poly(sulfone). Poly(sulfone) EFM decorated with MgO nanoparticles was found to be highly effective against paraoxon.

4.2.1.2 Enzymatic Degradation

Enzymatic degradation is the most selective and fastest approach for degradation of OPs. Enzymes result in efficient hydrolysis of OPs [52]. Han et al. [53] used coaxial electrospinning to fabricate composite EFM with a polymeric core and sheath of diisopropylfluorophosphatase (DFPase). DFPase is an enzyme able to efficiently hydrolyze OPs. Nevertheless, low stability and high production cost of enzymes drastically reduce its usage. Therefore, varieties of surrogate for enzymes have been used to replicate their effectiveness against OPs. For instance, cyclodextrin and its derivatives were used to mimic effectiveness of enzymes against OPs [54–57].

4.2.1.3 Photocatalytic Degradation

Heterogeneous photocatalysts have been shown to degrade OPs into harmless end products [58]. Typically, when a photocatalyst such as TiO_2 is irradiated with sunlight, photons are absorbed and electron–hole pairs are created. These electron–hole pairs participate in several redox reactions to generate reactive species responsible for the degradation of OPs. The most widely used photocatalyst TiO_2 has a bandgap ~3.2 eV, which allows it to absorb mainly ultraviolet light. Therefore, the structure/composition of TiO_2 is being modified to reduce its bandgap to improve its visible light photocatalytic activity. Further, work is in process to reduce its high recombination rate of electron–hole pairs. One of the highly applied techniques includes doping of TiO_2 with foreign ions to allow it to sensitize in visible light and also create charge traps for electron–hole pairs to keep them as separate identities [59].

TiO$_2$ has been doped with metals such as Mn, V, Fe, and Cu to improve its visible light sensitization. Nevertheless, large quantities of these dopants may increase recombination of electron–hole pairs. However, low concentration of dopants results in a small shift in the bandgap of TiO_2 toward visible light sensitization. Nonmetals such as N, S, and C are also used to reduce the bandgap of TiO_2. Thus, judiciously chosen low concentration of cation and anion co-dopants should enhance visible light sensitization and reduce recombination of electron–hole pairs [59]. Apart from TiO_2, other photocatalysts that can be used to degrade OPs include ZnO, SiO_2, and CuO.

4.2.1.4 Biological Warfare Agent Degradation

Biological warfare agents (*BWAs*) such as anthrax have acted as a threat for both military personnel and civilian population. Therefore, there are many research efforts to develop PT to prevent attack, and growth of BWAs. EFMs are functionalized with antibacterial and antiviral material to be used as PT. Typically, antibacterial materials used are silver, biguanide compounds, iodide complex, and phosphonium compounds, which can either be incorporated into or deposited onto EFMs to be effective against BWAs [60].

Studies have shown that EFMs containing antibacterial material such as Ag, MgO, and CuO are highly effective against bacterial threats [61–63]. These materials can either be blended with electrospinning solution or can be deposited on the surface of EFMs [63]. Benzyl triethylammonium chloride was blended with poly(carbonate) solution to fabricate an antibacterial EFM [64]. Photo-cross-linked chitosan/

Fig. 4.5 Schematic of smart protective textiles

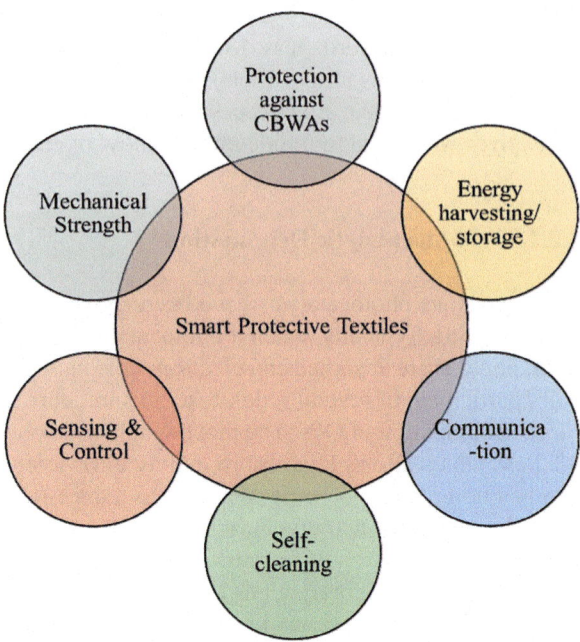

poly(vinyl alcohol) EFM was effective against *E. coli* and *S. aureus* [65]. Further, EFM containing chlorhexidine had also shown good antibacterial functionality [66].

4.3 Smart Protective Textiles

The future of smart PTs can be realized by incorporating functionalities such as sensing, energy harvesting/storage, communication, and self-cleaning (see Fig. 4.5). These smart PTs will not only provide protection against CBWAs but will also offer thermal comfort, man–machine interface, communication between personnel, and remote command center, self-cleaning capability, mechanical stability, and durability. Particularly, these smart PTs with sensing and communication capabilities can be remotely monitored for their effectiveness against CBWAs. Nevertheless, smart PTs can only be realized when collaboration between industry and academia is achieved to bridge the gap between research and mass-scale production.

4.3.1 Sensing and Control

Metal oxide sensors such as SnO_2, TiO_2, and In_2O_3 are widely used in textiles [67]. They can be readily attached to PTs for the detection of CBWAs [68]. CBWAs can be easily detected by measuring the variation in the electrical current or resistance when in contact with sensors. Nevertheless, metal oxide sensors operate at high temperature, usually above 300 °C with poor selectivity. Therefore, conducting polymers such as poly(aniline) and poly(ethylenedioxythiophene) are considered for sensing applications by virtue of their room temperature operation, and possibility of multi-functional derivatives [69]. Nevertheless, these sensors suffer from reduced sensitivity [69]. Therefore, novel sensors should consist of nanoscale organic/inorganic hybrid structures to achieve high selectivity with low response time. These new-generation sensors will selectively and rapidly detect CBWAs.

Further, smart gels can be employed to achieve control of temperature and permeability of PTs. For instance, the poly(N-isopropylacrylamide) hydrogel grafted onto poly(ethylene terephthalate) and poly(propylene) EFMs can change their volume depending on the operating temperature [70]. These hydrogel-grafted EFMs can be used to control the temperature of PTs. Further, poly(N-isopropylacrylamide-co-acrylamide)/chitosan microgels can be incorporated into EFM to yield dual temperature–responsive PTs to achieve thermal comfort for the wearer [71]. Further, poly(ethylene glycol) can be deposited on PTs for temperature control. In poly(ethylene glycol), hydrogen bond breaks and forms with an increase and decrease in the temperature, respectively, a property that can be exploited to sense and control the temperature of PTs [72].

4.3.2 Energy Harvesting/Storage and Communication

New-generation PTs will contain energy harvesting/storage and communication devices to measure and communicate vital human functions between the personnel and the command center for early detection and prevention of CBWAs. Therefore, PTs will potentially contain electro-active materials, which can be used in actuation, energy storage, and energy harvesting applications [73]. Typical electro-active materials include carbon nanotubes, conductive polymers, dielectric elastomers, and ferro-electronic polymers. Carbon nanotubes and their derivatives have been tested as actuators, which can potentially be used as an energy conversion device [74]. Conductive polymers such as poly(pyrrole), poly(aniline), and poly(thiophene) can also act as actuators. These polymers work on reversible counterion intake and removal during redox reactions [75]. Further, ionic membranes containing metals, such as Pt, Au, and Ag, have been positively tested as actuators [76].

4.3.3 Self-Cleaning

Dirt and fouling can degrade the performance of PTs due to blockage of pores preventing normal operation. Therefore, self-cleaning PTs are necessary for their effective performance over a long period of time. Typically, self-cleaning can be achieved via surface modification or photocatalytic degradation of organic dirt particles.

Surface modification makes the surface hydrophobic allowing the water droplet to roll off and pick up dusts, inorganic and organic contaminants along its way. Liu et al. [77] used surface-modified carbon nanotubes PBA-g-CNTs to fabricate lotus leaf–like hydrophobic structures on the surface of cotton fabric to generate super-hydrophobic textiles. Alternatively, organic contaminants can be decomposed via superoxide anions and hydroxyl radicals formed during a photocatalytic reaction in the presence of photocatalysts such as TiO_2 or ZnO and CuO [78, 79]. For example, cotton textile was surface-modified with TiO_2 nanoparticles to fabricate self-cleaning textiles [80]. Further, TiO_2 is combined with Au and SiO_2 to improve its photocatalytic activity and fabricate self-cleaning textile. Au reduces recombination between photo-generated electron–hole pairs, whereas amorphous SiO_2 increments photocatalytic activity of TiO_2 [81].

4.3.4   Mechanical Strength

Typical mechanical strength of EFMs is not adequate enough to withstand impacts during the flow of liquid or air through them. Therefore, EFMs are usually supported on a substrate or bonded with other fibrous media to enhance their mechanical strength. Novel approaches include the use of shape memory polymers in EFMs for PTs. Typical, shape memory polymers include poly(ethers), poly(amides), poly(acrylates), poly(siloxanes), and poly(urethanes). These shape memory polymers can reversibly deform under the influence of external forces such as temperature, pH, light, mechanical or chemical stimulus [74, 82]. These shape memory polymers will allow EFMs to deform reversibly under the action of liquid or gas flowing through it. This in turn will improve the mechanical stability of EFMs [83, 84].

4.4   Conclusion

This chapter discusses the use of EFM in protective textiles. Varieties of polymeric, ceramic, and composite single/multilayer EFM-based PTs have been fabricated and tested for their effectiveness against CBWAs. These lightweight EFM-based PTs are preferred by virtue of their high specific surface area, high breathability, and high porosity, while simultaneously providing the flexibility to incorporate required functionalities. Further, apart from adsorbing CBWAs, PTs are capable of decomposing them into nontoxic products, suggesting their feasibility for the new generation of PTs. Nevertheless, apart from protection against CBWAs, the new generation of PTs should also provide thermal comfort, communication between personnel, and remote command center, self-cleaning capability, sensing capability, as well as durability. Therefore, interdisciplinary research approaches are required to achieve a new generation of robust, sensitive, selective, and interactive PTs.

References

1. Scott RA (2005) Textiles for protection. Woodhead Pub, Sawston
2. Bromberg L, Schreuder-Gibson H, Creasy WR et al (2009) Degradation of chemical warfare agents by reactive polymers. Ind Eng Chem Res 48:1650–1659. https://doi.org/10.1021/ie801150y
3. Gugliuzza A, Drioli E (2013) A review on membrane engineering for innovation in wearable fabrics and protective textiles. J Membr Sci 446:350–375. https://doi.org/10.1016/j.memsci.2013.07.014
4. Dolez PI (2013) Smart barrier membranes for protective clothing. In: Smart textiles for protection. Elsevier, Amsterdam, pp 148–189
5. Raza A, Li Y, Sheng J et al (2014) Protective clothing based on electrospun nanofibrous membranes. In: Electrospun nanofibers for energy and environmental applications. Springer, Berlin, pp 355–369
6. Yoon B, Lee S (2011) Designing waterproof breathable materials based on electrospun nanofibers and assessing the performance characteristics. Fiber Polym 12:57–64. https://doi.org/10.1007/s12221-011-0057-9
7. Deitzel JM, Beck Tan NC, Kleinmeyer JD, et al (1999) Generation of polymer nanofibers through electrospinning. Army Res Rep ARL-TR-198:1–41
8. Online Army Study Guide - Chemical, Biological, Radiological, Nuclear | ArmyStudyGuide. com. http://www.armystudyguide.com/content/army_board_study_guide_topics/cbrn/cbrn-study-guide.shtml. Accessed 19 Jul 2017
9. Sahay R, Teo CJ, Chew YT (2013) New correlation formulae for the straight section of the electrospun jet from a polymer drop. J Fluid Mech 735:150–175. https://doi.org/10.1017/Jfm.2013.497
10. Sahay R, Thavasi V, Ramakrishna S (2011) Design modifications in electrospinning setup for advanced applications. J Nanomater 2011:1–17. https://doi.org/10.1155/2011/317673
11. Sahay R, Kumar PS, Sridhar R et al (2012) Electrospun composite nanofibers and their multifaceted applications. J Mater Chem 22:12953. https://doi.org/10.1039/c2jm30966a
12. Gibson P, Schreuder-Gibson H, Rivin D (2001) Transport properties of porous membranes based on electrospun nanofibers. Colloids Surf A Physicochem Eng Asp 187–188:469–481. https://doi.org/10.1016/S0927-7757(01)00616-1
13. Lee S, Obendorf SK (2007) Use of electrospun nanofiber web for protective textile materials as barriers to liquid penetration. Text Res J 77:696–702. https://doi.org/10.1177/0040517507080284
14. Ecobichon DJ (1999) Occupational hazards of pesticide exposure: sampling, monitoring, and measuring. Taylor & Francis, Abingdon
15. Davies HG, Richter RJ, Keifer M et al (1996) The effect of the human serum paraoxonase polymorphism is reversed with diazoxon, soman and sarin. Nat Genet 14:334–336. https://doi.org/10.1038/ng1196-334
16. Taylor P (2010) The degradation of organophosphorus pesticides in natural waters: a the degradation of organophosphorus pesticides in natural waters: a critical review. Crit Rev Environ Sci Technol 32:17–72. https://doi.org/10.1080/10643380290813444
17. Singh BK (2009) Organophosphorus-degrading bacteria: ecology and industrial applications. Nat Rev Microbiol 7:156–164. https://doi.org/10.1038/nrmicro2050
18. Voss G, Matsumura F (1964) Resistance to organophosphorus compounds in the two-spotted spider mite: two different mechanisms of resistance. Nature 202:319–320
19. Yang Y, Baker J, Ward J (1992) Decontamination of chemical warfare agents. Chem Rev 92:1729–1743
20. Mondloch JE, Katz MJ, Isley WC III et al (2015) Destruction of chemical warfare agents using metal–organic frameworks. Nat Mater 14:512–516. https://doi.org/10.1038/nmat4238

21. Borak J, Sidell FR (1992) Agents of chemical warfare: sulfur mustard. Ann Emerg Med 21:303–308. https://doi.org/10.1016/S0196-0644(05)80892-3
22. Eddleston M, Buckley NA, Eyer P, Dawson AH (2008) Management of acute organophosphorus pesticide poisoning. Lancet 371:597–607. https://doi.org/10.1016/S0140-6736(07)61202-1
23. Senić Ž, Bauk S, Vitorović-Todorović M et al (2011) Application of TiO$_2$ nanoparticles for obtaining self-decontaminating smart textiles. Sci Tech Rev 61:63–72
24. Ultra-Web Media Technology - Dust Collector Filters - Donaldson Torit. http://www2.donaldson.com/torit/en-us/pages/products/ultra-webmediatechnology.aspx. Accessed 23 Jul 2017
25. Brock R, Meitner G (1977) Nonwoven thermoplastic fabric. US Pat. 4,041,203
26. Groitzsch D, Fahrbach E (1986) Microporous multilayer nonwoven material for medical applications. US Pat. 4,618,524
27. Huber O, Magidson M (1983) Disposable face mask. US Pat. 4,384,577
28. Saint-André G, Kliachyna M, Kodepelly S et al (2011) Design, synthesis and evaluation of new α-nucleophiles for the hydrolysis of organophosphorus nerve agents: application to the reactivation of phosphorylated acetylcholinesterase. Tetrahedron 67:6352–6361. https://doi.org/10.1016/j.tet.2011.05.130
29. Faust SD, Gomaa HM (1972) Chemical hydrolysis of some organic phosphorus and carbamate pesticides in aquatic environments. Environ Lett 3:171–201. https://doi.org/10.1080/00139307209435465
30. Fukuto TR (1990) Mechanism of action of organophosphorus and carbamate insecticides. Environ Health Perspect 87:245–254
31. Mutin PH, Guerrero G, Vioux A (2005) Hybrid materials from organophosphorus coupling molecules. J Mater Chem 15:3761–3768. https://doi.org/10.1039/b505422b
32. Stark JV, Park DG, Lagadic I, Klabunde KJ (1996) Nanoscale metal oxide particles/clusters as chemical reagents. Unique surface chemistry on magnesium oxide as shown by enhanced adsorption of acid gases (sulfur dioxide and carbon dioxide) and pressure dependence. Chem Mater 8:1904–1912. https://doi.org/10.1021/cm950583p
33. Li YX, Koper O, Atteya M, Klabunde KJ (1992) Adsorption and decomposition of organophosphorus compounds on nanoscale metal oxide particles. In situ GC-MS studies of pulsed microreactions over magnesium oxide. Chem Mater 4:323–330. https://doi.org/10.1021/cm00020a019
34. Bunton C, Foroudian HJ, Gillitt ND (1999) Reduction ofo-iodosobenzoate ion by sulfides and its oxidative regeneration. J Phys Org Chem 12:758–764. https://doi.org/10.1002/(SICI)1099-1395(199910)12:10<758::AID-POC200>3.0.CO;2-A
35. Graham K, Schreuder-Gibson H, Gogins M (2003) Incorporation of electrospun nanofibers into functional structures. Construction 15–18
36. Chen L, Bromberg L, Lee JA et al (2010) Multifunctional electrospun fabrics via layer-by-layer electrostatic assembly for chemical and biological protection. Chem Mater 22:1429–1436. https://doi.org/10.1021/cm902834a
37. Ramakrishna S, Fujihara K, Teo WE et al (2006) Electrospun nanofibers: solving global issues. Mater Today 9:40–50. https://doi.org/10.1016/S1369-7021(06)71389-X
38. Ramaseshan R, Sundarrajan S, Liu Y et al (2006) Functionalized polymer nanofibre membranes for protection from chemical warfare stimulants. Nanotechnology 17:2947–2953. https://doi.org/10.1088/0957-4484/17/12/021
39. Ramaseshan R (2011) Decontamination of chemical warfare simulants using electrospun media. PhD Thesis, National University Singapore, pp 1–167
40. Chen L, Bromberg L, Hatton TA, Rutledge GC (2007) Catalytic hydrolysis of p-nitrophenyl acetate by electrospun polyacrylamidoxime nanofibers. Polymer 48:4675–4682. https://doi.org/10.1016/j.polymer.2007.05.084
41. Chen L, Bromberg L, Schreuder-Gibson H et al (2009) Chemical protection fabrics via surface oximation of electrospun polyacrylonitrile fiber mats. J Mater Chem 19:2432–2438. https://doi.org/10.1039/B818639a

42. Becke GS, Carmody DJ, Dobosy MJ (2008) Microporous breathable film with internal barrier layer or layers. US20080131676 A1
43. Nooney MG, Campbell A, Murrell TS et al (1998) Nucleation and growth of phosphate on metal oxide thin films. Langmuir 14:2750–2755. https://doi.org/10.1021/la9702695
44. Li M, Liu J, Xu Y, Qian G (2016) Phosphate adsorption on metal oxides and metal hydroxides: a comparative review. Environ Rev 24:319–332. https://doi.org/10.1139/er-2015-0080
45. Wagner GW, Bartram PW, Koper O, Klabunde KJ (1999) Reactions of VX, GD, and HD with Nanosize MgO. J Phys Chem B 103:3225–3228. https://doi.org/10.1021/jp984689u
46. Kleinhammes A, Wagner GW, Kulkarni H et al (2005) Decontamination of 2-chloroethyl ethylsulfide using titanate nanoscrolls. Chem Phys Lett 411:81–85. https://doi.org/10.1016/j.cplett.2005.05.100
47. Šťastný M, Štengl V, Henych J et al (2016) Mesoporous manganese oxide for the degradation of organophosphates pesticides. J Mater Sci 51:2634–2642. https://doi.org/10.1007/s10853-015-9577-9
48. Bootharaju MS, Pradeep T (2012) Understanding the degradation pathway of the pesticide, chlorpyrifos by noble metal nanoparticles. Langmuir 28:2671–2679. https://doi.org/10.1021/la2050515
49. Silva VB, Rodrigues TS, Camargo PHC, Orth ES (2017) Detoxification of organophosphates using imidazole-coated Ag, Au and AgAu nanoparticles. RSC Adv 7:40711–40719. https://doi.org/10.1039/c7ra07059d
50. Ramaseshan R, Ramakrishna S (2007) Zinc titanate nanofibers for the detoxification of chemical warfare simulants. J Am Ceram Soc 90:1836–1842. https://doi.org/10.1111/j.1551-2916.2007.01633.x
51. Sundarrajan S, Ramakrishna S (2007) Fabrication of nanocomposite membranes from nanofibers and nanoparticles for protection against chemical warfare stimulants. J Mater Sci 42:8400–8407. https://doi.org/10.1007/s10853-007-1786-4
52. Aubert SD, Li Y, Raushel FM (2004) Mechanism for the hydrolysis of organophosphates by the bacterial phosphotriesterase. Biochemistry 43:5707–5715. https://doi.org/10.1021/bi0497805
53. Han D, Filocamo S, Kirby R, Steckl AJ (2011) Deactivating chemical agents using enzyme-coated nanofibers formed by electrospinning. ACS Appl Mater Interfaces 3:4633–4639. https://doi.org/10.1021/am201064b
54. Moss RA, Alwis KW, Bizzigotti GO (1983) o-Iodosobenzoate: catalyst for the micellar cleavage of activated esters and phosphates. J Am Chem Soc 105:681–682. https://doi.org/10.1021/ja00341a092
55. Moss RA, Alwis KW, Shin JS (1984) Catalytic cleavage of active phosphate and ester substrates by iodoso- and iodoxybenzoates. J Am Chem Soc 106:2651–2655. https://doi.org/10.1021/ja00321a027
56. Menger FM, Rourk MJ (1999) Deactivation of mustard and nerve agent models via low-temperature microemulsions. Langmuir 15:309–313. https://doi.org/10.1021/la980910i
57. Li Y-F, Ha Y-M, Guo Q, Li Q-P (2015) Synthesis of two β-cyclodextrin derivatives containing a vinyl group. Carbohydr Res 404:55–62. https://doi.org/10.1016/j.carres.2014.11.012
58. Echavia GRM, Matzusawa F, Negishi N (2009) Photocatalytic degradation of organophosphate and phosphonoglycine pesticides using TiO_2 immobilized on silica gel. Chemosphere 76:595–600. https://doi.org/10.1016/j.chemosphere.2009.04.055
59. Di Valentin C, Pacchioni G, Selloni A (2009) Reduced and n-type doped TiO_2: nature of Ti^{3+} species. J Phys Chem C 113:20543–20552. https://doi.org/10.1021/jp9061797
60. Hong KH, Park JL, Hwan Sul IN et al (2006) Preparation of antimicrobial poly(vinyl alcohol) nanofibers containing silver nanoparticles. J Polym Sci B Polym Phys 44:2468–2474. https://doi.org/10.1002/polb.20913
61. Duan YY, Jia J, Wang SH et al (2007) Preparation of antimicrobial poly(e-caprolactone) electrospun nanofibers containing silver-loaded zirconium phosphate nanoparticles. J Appl Polym Sci 106:1208–1214. https://doi.org/10.1002/app.26786

62. Haider A, Kwak S, Gupta KC, Kang IK (2015) Antibacterial activity and cytocompatibility of PLGA/CuO hybrid nanofiber scaffolds prepared by electrospinning. J Nanomater 2015:1–10. https://doi.org/10.1155/2015/832762

63. Yuan J, Geng J, Xing Z et al (2010) Electrospinning of antibacterial poly(vinylidene fluoride) nanofibers containing silver nanoparticles. J Appl Polym Sci 116:668–672. https://doi.org/10.1002/app.31632

64. Kim SJ, Nam YS, Rhee DM et al (2007) Preparation and characterization of antimicrobial polycarbonate nanofibrous membrane. Eur Polym J 43:3146–3152. https://doi.org/10.1016/j.eurpolymj.2007.04.046

65. Ignatova M, Starbova K, Markova N et al (2006) Electrospun nano-fibre mats with antibacterial properties from quaternised chitosan and poly(vinyl alcohol). Carbohydr Res 341:2098–2107. https://doi.org/10.1016/j.carres.2006.05.006

66. Chen L (2009) Next generation of electrospun textiles for chemical and biological protection and air filtration. PhD Thesis, Massachusetts Institute of Technology, pp 1–165

67. Kent JA (2012) Handbook of industrial chemistry and biotechnology. Springer, Berlin, pp 1–1562

68. Kiekens P, Jayaraman S (2012) Intelligent textiles and clothing for ballistic and NBC protection: technology at the cutting edge. Springer, Berlin, pp 1–220

69. Rothschild A, Komem Y (2003) Numerical computation of chemisorption isotherms for device modeling of semiconductor gas sensors. Sensors Actuators B Chem 93:362–369. https://doi.org/10.1016/S0925-4005(03)00212-0

70. Wang J, Zhong Q, Wu J, Chen T (2014) Thermo-responsive textiles. In: Handbook of smart textiles. Springer, Singapore, pp 1–27

71. Križman Lavrič P, Warmoeskerken MMCG, Jocic D (2012) Functionalization of cotton with poly-NiPAAm/chitosan microgel. Part I. Stimuli-responsive moisture management properties. Cellulose 19:257–271. https://doi.org/10.1007/s10570-011-9632-x

72. Meldrum FC (2005) Biomineralisation processes. In: Surfaces interfaces biomater, pp 666–692. https://doi.org/10.1533/9781845690809.4.666

73. Safadi B, Andrews R, Grulke EA (2002) Multiwalled carbon nanotube polymer composites: synthesis and characterization of thin films. J Appl Polym Sci 84:2660–2669. https://doi.org/10.1002/app.10436

74. Hu J, Zhu Y, Huang H, Lu J (2012) Recent advances in shape–memory polymers: structure, mechanism, functionality, modeling and applications. Prog Polym Sci 37:1720–1763. https://doi.org/10.1016/j.progpolymsci.2012.06.001

75. Uchino K (2010) Advanced piezoelectric materials science and technology. Woodhead Publ, Sawston, pp 1–678

76. Fink JK (2012) Polymeric sensors and actuators. Wiley, Hoboken, pp 1–512

77. Liu YY, Wang RH, Lu HF et al (2007) Artificial lotus leaf structures from assembling carbon nanotubes and their applications in hydrophobic textiles. J Mater Chem 17:1071–1078. https://doi.org/10.1039/B613914k

78. Banerjee S, Dionysiou D, Pillai S (2015) Self-cleaning applications of TiO_2 by photo-induced hydrophilicity and photocatalysis. Appl Catal B Environ 176:396–428. https://doi.org/10.1016/j.apcatb.2015.03.058

79. Momeni MM, Ghayeb Y, Ghonchegi Z (2015) Fabrication and characterization of copper doped TiO_2 nanotube arrays by in situ electrochemical method as efficient visible-light photocatalyst. Ceram Int 41:8735–8741. https://doi.org/10.1016/j.ceramint.2015.03.094

80. Wu D, Long M, Zhou J et al (2009) Synthesis and characterization of self-cleaning cotton fabrics modified by TiO_2 through a facile approach. Surf Coat Technol 203:3728–3733. https://doi.org/10.1016/j.surfcoat.2009.06.008

81. Kaihong Qi K, Xiaowen Wang X, Xin JH (2011) Photocatalytic self-cleaning textiles based on nanocrystalline titanium dioxide. Text Res J 81:101–110. https://doi.org/10.1177/0040517510383618

82. Liu C, Qin H, Mather PT et al (2007) Review of progress in shape-memory polymers. J Mater Chem 17:1543. https://doi.org/10.1039/b615954k
83. Han HR, Chung SE, Park CH (2013) Shape memory and breathable waterproof properties of polyurethane nanowebs. Text Res J 83:76–82. https://doi.org/10.1177/0040517512450757
84. Pretsch T (2010) Review on the functional determinants and durability of shape memory polymers. Polymers 2:120–158. https://doi.org/10.3390/polym2030120

Chapter 5
Electrospun Filters for Heavy Metals Removal

Rui Zhao, Xiang Li, and Ce Wang

Abstract Due to high porosity, large surface-to-volume ratio, film characteristics, and interconnectivity of electrospun fibers, they have shown great potential in wastewater treatment, especially in the heavy metals removal field. The adsorption and filtration are the most common methods to treat the heavy metal-polluted wastewater. Possessing the above characteristics makes electrospun fibers being suitable adsorbents or filters. Adsorption is a relatively static process where adsorbates are adsorbed by the adsorbent via physical and chemical interactions in the contaminated solution. The filtration is a dynamic process that involves the polluted wastewater passing through the filters by its own weight or other forces, to separate the pollutants from the wastewater. However, the filtration of heavy metal ions by the electrospun filters is mainly due to the adsorption processes between the filters and the metal ions, and the electrospun adsorbents are widely studied for the heavy metal remediation. Thus, this chapter's scope focuses on two areas (electrospun adsorbents and filters for heavy metals removal) and gives representative examples of electrospun adsorbents or filters that are used for the removal of heavy metals.

5.1 Introduction

The rapid growth of urbanization and industrialization, such as metal plating facilities, fertilizer industries, mining, tanneries, batteries, car manufacturing, pesticides, etc., has resulted in the increase of heavy metals-contaminated wastewater discharge into the environment around the world [1–3]. According to a recent research report, the water resource scarcity will be a greater challenge toward 2030 due to the huge pressure on limited available fresh water sources [4, 5]. As a kind of important pollutants, heavy metals are difficult to be degraded and they tend to accumulate in living organisms [6]. Usual discharged heavy metals in waters are As, Pb, Hg, Cd, Cr, Cu, Ni, Zn, and so on. Although trace amounts of different heavy metals are necessary for human beings, excessive hazardous heavy metals in water are a risk

R. Zhao · X. Li · C. Wang (✉)
Alan G. MacDiarmid Institute, College of Chemistry, Jilin University, Changchun, PR China
e-mail: cwang@jlu.edu.cn

© Springer International Publishing AG, part of Springer Nature 2018
M. L. Focarete et al. (eds.), *Filtering Media by Electrospinning*,
https://doi.org/10.1007/978-3-319-78163-1_5

and a significant environmental problem throughout the world. Moreover, most heavy metals are highly toxic or carcinogenic. In this matter, various conventional and modern methods, such as chemical precipitation, ion exchange, adsorption, membrane filtration, electrochemical treatment, coagulation and flocculation, flotation, etc., have been used to treat the heavy metal-contaminated water in the past few decades [7–10]. Adsorption and filtration are among the most preferred methods for the removal of heavy metal ions from wastewater because of their advantages of easy operation, low cost, and high efficiency [11–13]. Adsorption is a process in which adsorbates are adsorbed by the adsorbent via physical and chemical interactions in the contaminated solution. The adsorption process is easy; moreover, this process is mostly reversible and the adsorbents can be regenerated through a suitable desorption process. The filtration process involves the polluted wastewater passing through the filters by its own weight or other forces, to separate the pollutants from the wastewater. Compared with dynamic filtration, adsorption is a relatively static process. For adsorption or filtration processes, removal selectivity and capacity are two of the most important characteristics of the adsorbents or filters, which depend on their physical and chemistry properties. Many materials have been applied for these two processes to remove heavy metal ions. Among these materials, electrospun one-dimensional fibrous materials have received considerable attention in the two fields in recent years [14–16].

Electrospinning is a facile, effective, low-cost, and highly versatile approach to fabricate one-dimensional continuous fibers with diameters down to nanometer range under a high electric field. The first description of an apparatus for this electrospinning technique was demonstrated by Formhals in 1934 as a patent; however, until the early of 1990s, this technique did not receive much attention [17]. Then, Reneker and coworkers made a great contribution to the development of this technique [18–22]. The typical electrospinning setup contains three major components: high-voltage power supply, spinneret, and collector. During the electrospinning process, a high electric field is applied between the droplet of a fluid (polymer solution/melt) coming out from the spinneret and the collector. Once the strength of the electric field becomes sufficiently greater than the surface tension of the polymer droplet, a charged liquid jet is formed and finally elongates to form a Taylor cone. Under the effect of electric field, the jet becomes accelerated toward the collector, leading to the formation of continuous fibers [23, 24].

Electrospun fibers have shown a broad range of applications, owing to their low dimensions and large surface areas, such as templates for fabricating other nanostructures, filters, textiles, catalysis, sensors, scaffolds in medicine, energy materials, and so on [25–28]. One of the advantages of the electrospinning technique is the flexible design of the setup to prepare the product with specific composition and structure [23, 25]. Moreover, electrospun fibers show high porosity, large surface-to-volume ratio, easy preparation, good membrane formation and flexibility, making them suitable in adsorption and filtration fields [15, 16]. In the past few years, increasing attention has been given to the electrospun fibers for heavy metals removal [29]. For the electrospun fiber materials, the removal mechanism between the fiber and the heavy metals is mainly the chemical interaction (such as chelating and electrostatic interaction). In addition, the physical interaction is also a determining factor [14, 29]. In order to remove heavy metals, researchers designed different types of electrospun fibers, including inorganic nanofibers, organic nanofibers, and organic/inorganic composite nanofibers with different structures. The adsorption and filtration are the most common methods to treat the heavy metal wastewater. The ion radius of heavy metal ions is very small and the filtration of the heavy metal by the electrospun fiber membrane is mainly dependent on the chemical adsorption process. Thus, most of the researches aim at the static adsorption which is the foundation for the heavy metal filtration and the dynamic filtration research is limited. This chapter contains two main sections: electrospun adsorbents for static heavy metals removal and electrospun filters for dynamic heavy metals removal. In particular, electrospun heavy metal adsorbents are presented in detail.

5.2 Electrospun Adsorbents for Heavy Metals Removal

Due to the characteristics mentioned above (large surface-to-volume ratio, easy preparation, good membrane formation, and flexible design), different types of electrospun fibers, including inorganic nanofibers, organic nanofibers, and organic/inorganic composite nanofibers with different structures are prepared to adsorb heavy metal ions in the wastewater.

5.2.1 Inorganic Electrospun Nanofibers

The most common method for the preparation of inorganic electrospun nanofibers is the combination of electrospinning and sol–gel techniques. Inorganic nanofibers are obtained by the following procedure: (1) preparation of a spinnable solution containing a sol–gel precursor, (2) electrospinning of the sol–gel solution, and (3) calcination of the as-spun nanofibers in air removing organic templates to obtain inorganic nanofibers. Inorganic nanofibers have high surface area and specific affinity for the adsorption; moreover, electrospinning technology shows low cost and high yield [27, 30]. Many types of inorganic electrospun nanofibers, such as one-component and multicomponent inorganic nanofibers, have been used to adsorb heavy metal ions.

The nano-sized metal oxide is well known to show high uptake of cations and anions in natural environments due its good physical and chemical properties. $\alpha\text{-}Fe_2O_3$ is considered as an adsorbent for the removal of harmful heavy metals in water sources. In the work of Myung's group, electrospinning provided a simple synthesis route to create $\alpha\text{-}Fe_2O_3$ nanofibers with different average diameters ranging from 23 to 63 nm [31]. Due to the increased surface area-to-volume ratio with decreasing nanofiber size, the specific surface area of the $\alpha\text{-}Fe_2O_3$ nanofibers increased with the decreasing of average diameter, from 7.2 to 59.2 m^2/g. The

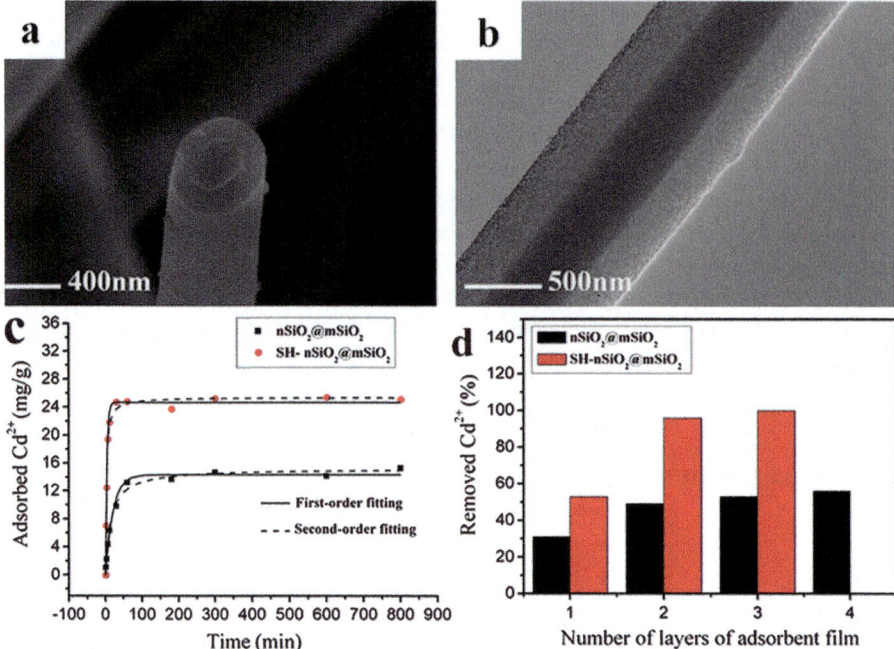

Fig. 5.1 High magnified SEM image (**a**) and TEM image (**b**) of one single nSiO$_2$@mSiO$_2$ fiber. (**c**) Adsorption kinetics Cd^{2+} adsorption by nSiO$_2$@mSiO$_2$ and thiol grafted nSiO$_2$@mSiO$_2$. (**d**) Adsorption performance of films patterned from nSiO$_2$@mSiO$_2$ and thiol grafted nSiO$_2$@mSiO$_2$ for Cd^{2+}. Adapted from Ref. [32] and reprinted with permission from Elsevier

obtained α-Fe$_2$O$_3$ nanofibers were used to adsorb CrO$_4$$^{2-}$. According to the Langmuir fitted isotherms, the smallest Fe$_2$O$_3$ nanofibers with the diameter of 23 nm had the greatest adsorption capacity of 90.9 mg/g, which was two times larger than the commercially available Fe$_2$O$_3$ nanoparticles. In order to improve the adsorption capacity toward heavy metal ions, the inorganic electrospun nanofibers were also modified by a functional process. Ma and coworkers reported a large-scale fabrication of mesoporous SiO$_2$ with high specific surface area, high flexibility, and long fibrous morphology [32]. As shown in Fig. 5.1a and b, the SiO$_2$ fibers are composed of a nonporous core of SiO$_2$ nanofibers and a mesoporous shell (nSiO$_2$@mSiO$_2$). The flexible nonporous SiO$_2$ nanofiber was prepared with electrospinning, followed by covering a mesoporous SiO$_2$ shell based on a modified Stöber method. To improve the adsorption performance, (3-mercaptopropyl)trimethoxysilane was used as coupling agent to graft functional thiol groups on the surface of nSiO$_2$@mSiO$_2$. After the grafting of thiol groups, adsorption rate and capacity were both significantly enhanced for Cd^{2+} ions. The equilibrium time decreased to 10 min from 30 min and the saturated adsorption capacity increased to 30.22 mg/g from 18.09 mg/g (Fig. 5.1c). The authors also assembled the nSiO$_2$@mSiO$_2$ and grafted nSiO$_2$@mSiO$_2$ into filtration films. For Cd^{2+}, after passing through four layers of nSiO$_2$@

mSiO$_2$ films, 44% Cd^{2+} ions remained in the filtrate, while 96% Cd^{2+} was removed by only three layers of thiol-functionalized nSiO$_2$@mSiO$_2$ (Fig. 5.1d).

In addition to the metal oxide electrospun nanofibers, other types of inorganic nanofibers with special structures have also been prepared. Jin et al. used the electrospinning technique and hydrothermal method to synthesize the magnesium silicate double-walled hollow nanofibers (MSHNFs) with a hierarchical nanostructure [33]. Magnesium silicate materials are attractive as adsorbents owing to their low cost, environmental friendliness, and interesting structures. The MSHNFs were synthesized by a simple hydrothermal treatment by using electrospun SiO$_2$ hollow nanofibers (SHNFs) as a sacrificial template. SHNFs were dissolved to form silicate anions under alkaline conditions during the hydrothermal process, and then they reacted with the magnesium ions to form magnesium silicate, followed by depositing on both sides of the wall of the SHNFs. The SHNF wall could be dissolved completely to form the second cavity as the hydrothermal process continues. The prepared MSHNFs had a high specific surface area (632.2 m^2/g) and large pore volume (0.92 cm^3/g). The MSHNFs showed a high adsorption capacity of 158 mg/g toward Pb^{2+} through ion exchange. Significantly, the as-prepared MSHNFs could be easily separated by gravitational sedimentation. The above results showed that the MSHNFs were promising materials for water purification application. In another work of Du's group, porous silica nanotubes (SNTs) with controlled nanotubular structure (Fig. 5.2) were synthesized via an electrospinning and calcination process by altering the content of TEOS in the precursor and controlling appropriate

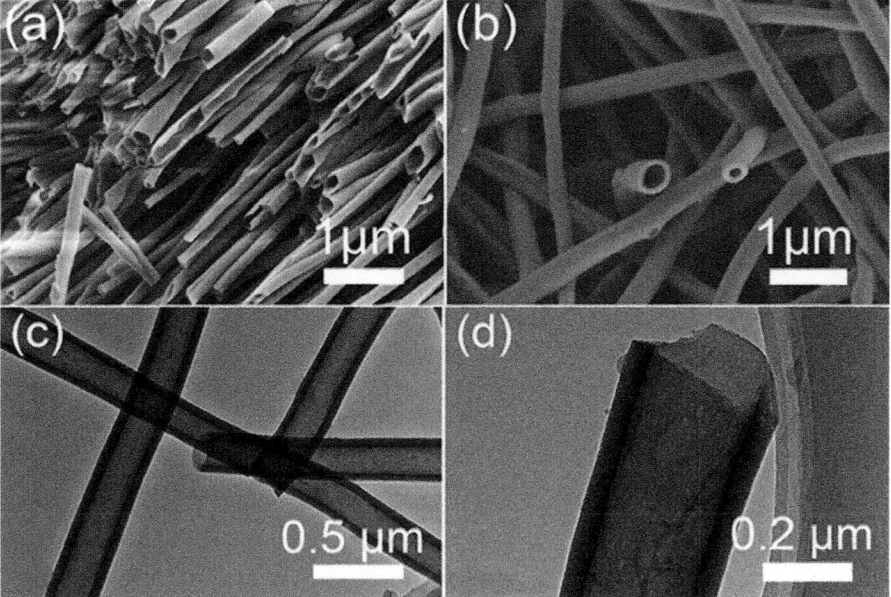

Fig. 5.2 SEM (**a, b**) and TEM (**c, d**) images of silica nanotubes. Adapted from Ref. [34] and reprinted with permission from Elsevier

calcination rates [34]. The obtained silica nanotubes were ideal adsorbents for Pb^{2+} removal; moreover, the surface of SNTs was modified by sym-diphenylcarbazide to further enhance the adsorption ability. The adsorption performance of SNTs and modified SNTs toward Pb^{2+} were discussed in detail, including pH value effects, kinetic studies, adsorption isotherms, and adsorption mechanism. The as-prepared SNTs and modified SNTs also showed high adsorption ability for Cd^{2+} and Co^{2+}, indicating their excellent ability to treat the wastewater contaminated with heavy metal ions.

Comparing with the single inorganic electrospun nanofibers, there is great interest in the synthesis of mixed metal oxide nanofibers and inorganic/carbon materials composite nanofibers. The adsorption capacity would be improved by the mixed or composite nanofibers due to the synergistic effects of the two or more components. Malwal et al. synthesized CuO–ZnO composite nanofibers by a simple electrospinning technique and post-calcination [35]. The as-prepared composite nanofibers were used to adsorb As from water. It was found that CuO–ZnO composite nanofibers possessed a fast adsorption rate and high adsorption capacity toward As in comparison with pure ZnO nanofibers. It was also observed that the adsorption data of As onto CuO–ZnO composite nanofibers fit both well with Langmuir and Freundlich isotherm models. The maximum adsorption capacity was 27.7 mg/g on the basis of the Langmuir adsorption isotherm model. The mixture is a simple way to prepare mixed metal oxide nanofibers, and metal oxide nanofibers with special architectures are also fabricated. Li and coworkers reported α-Fe_2O_3–γ-Al_2O_3 core–shell nanofibers which were synthesized by an electrospinning process combined with vapor deposition and heat treatment techniques [36]. The composite nanofibers showed ferromagnetic properties and were employed to remove Cr(VI) ions. The adsorption process followed the pseudo-second-order rate equation and the Freundlich isotherm model. The adsorption mechanism of Cr(VI) onto α-Fe_2O_3–γ--Al_2O_3 attributed to the electrostatic adsorption between the positively charged surface of α-Fe_2O_3–γ-Al_2O_3 nanofibers and Cr(VI) species, and the electron–hole pair provided by Fe_2O_3 induced the Cr(VI) reduction to Cr(III).

Carbon is considered as a cheap and common material for adsorption applications. In this regard, inorganic/carbon materials composite nanofibers are also studied for the removal of heavy metals. The pure inorganic nanofibers were obtained by calcining the inorganic salt/polymer composite fibers in the air atmosphere to remove the organic templates. Carbon nanofibers from the electrospun fibers were prepared by calcining the polymer fibers in the inert gas atmosphere via the carbonization process. Han et al. reported a facile method to prepare [C/Fe_3O_4]@C coaxial nanocables by electrospinning and carbonization process [37]. The schematic preparation process is shown in Fig. 5.3a. The coaxial nanocables showed tunable electrical conductivity and magnetic performance by altering Fe_3O_4 content. The [C/Fe_3O_4]@C coaxial nanocables exhibited good adsorption toward Cu^{2+} due to the Fe_3O_4, C components and their porous structures. The maximum adsorption capacity for Cu^{2+} was 64.06 mg/g (Fig. 5.3c). Due to the magnetic performance, the adsorbent could be recycled easily. In another study, ZrO_2 embedded in carbon nanowires (ZCNs) were fabricated by calcining the electrospun PVP/$ZrOCl_2$

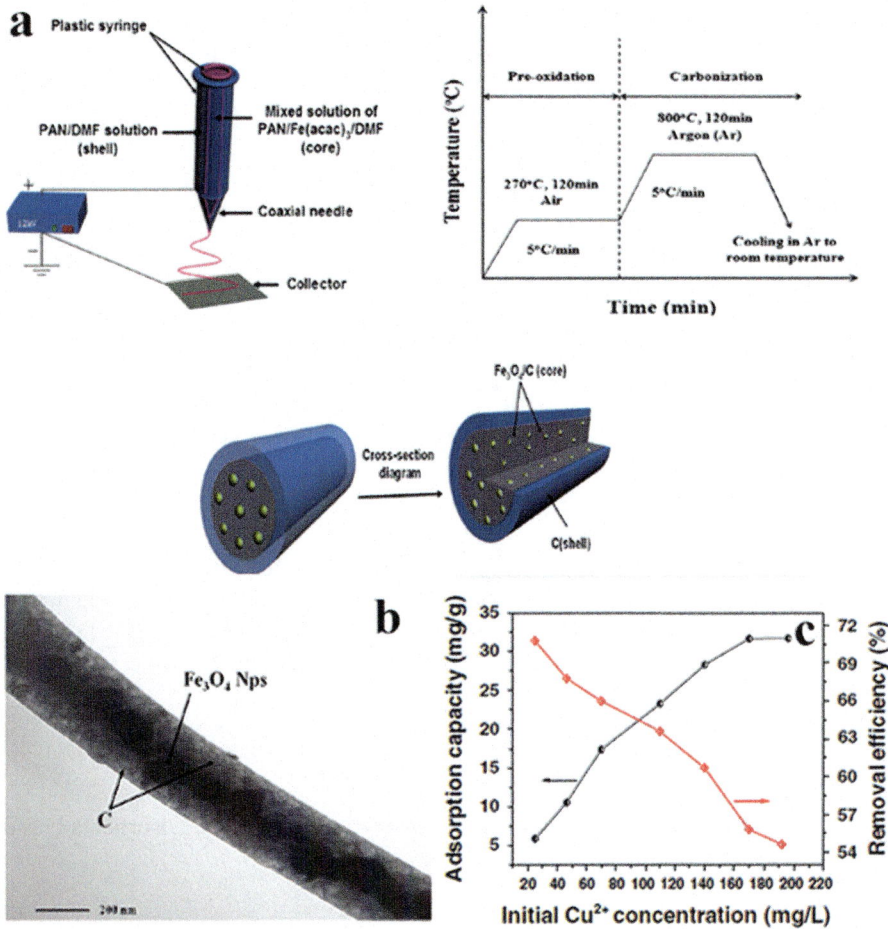

Fig. 5.3 (**a**) Schematic diagram of the synthesis of [C/Fe₃O₄]@C coaxial nanocables. (**b**) TEM image of [C/Fe₃O₄]@C coaxial nanocables. (**c**) Effect of initial Cu²⁺ concentration on the adsorption capacity and removal percentage. Adapted from Ref. [37] and reprinted with permission from Springer

composite nanowires under nitrogen (N₂) [38]. The carbon nanowires were employed as the support to improve its physical properties and lower the overall cost. The adsorption performance of arsenic (As) onto ZCNs was investigated. A wide pH range for As(III) (5.0–11.0) and As(V) (5.0–9.0) adsorption was achieved. The maximum adsorption capacity values of As(III) and As(V) on ZCNs were 28.61 and 106.57 mg/g at 40 °C, which were higher than those of pure ZrO₂ nanofibers, suggesting the synergistic effects of the two components. The adsorption thermodynamics and adsorption isotherm indicated that the adsorption processes were endothermic and followed the Freundlich isotherm model. In addition to the examples as aforementioned, there are several other inorganic electrospun nanofibers for heavy

Table 5.1 Inorganic electrospun nanofibers as heavy metal adsorbents

Adsorbent	Heavy metal	Optimum pH	Adsorption capacity (mg/g)	Reference
α-Fe$_2$O$_3$ nanofibers	CrO$_4^{2-}$	6	90.9	[31]
Anatase mesoporous TiO$_2$ nanofibers	Cu^{2+}	6	12.8	[39]
Electrospun meso-hydroxylapatite nanofibers	Co^{2+}	6.5	8.14	[40]
MgSi double-walled hollow nanofibers	Pb^{2+}	–	158	[33]
Thiol-modified nonporous SiO$_2$@ mesoporous SiO$_2$ fibers	Cd^{2+}	–	30.22	[32]
	Pb^{2+}	–	154.7	
Sym-diphenylcarbazide-modified silica nanotubes	Pb^{2+}	7	112.36	[34]
Zonal thiol-functionalized silica nanofibers	Hg^{2+}	–	57.49	[41]
α-Fe$_2$O$_3$–γ-Al$_2$O$_3$ core–shell nanofibers	CrO$_4^{2-}$	2.0	57.34	[36]
Electrospun Fe$_2$O$_3$–Al$_2$O$_3$ nanocomposite fibers	CrO$_4^{2-}$	6	4.98	[42]
	Pb^{2+}	6	23.75	
	Ni^{2+}	6	32.36	
	Hg^{2+}	6	63.69	
CuO–ZnO composite nanofibers	AsO$_2^-$	4-10	27.7	[35]
ZrO$_2$ embedded in carbon nanowires	AsO$_2^-$	5–11	28.61	[38]
	HAsO$_4^{2-}$	5–9	106.57	
Mesoporous hydroxylapatite/activated carbon bead-on-string nanofibers	Co^{2+}	6.5	8.26	[43]
[C/Fe$_3$O$_4$]@C coaxial nanocables	Cu^{2+}	6	64.06	[37]

metal adsorption. Table 5.1 summarized a variety of heavy metal adsorbents based on inorganic electrospun nanofibers.

5.2.2 Organic Electrospun Nanofibers

Although inorganic electrospun nanofibers showed good adsorption capacities toward heavy metals, the mechanical strength of inorganic nanofibers is limited and most of the inorganic electrospun nanofibers are in the powder form, restricting their practical applications. Electrospun polymer nanofibers display good film-forming properties and satisfying mechanical properties. Moreover, most of the polymers have functional groups which could bind with metal ions. Many synthetic polymers and natural polymers have been electrospun into heavy metal fiber adsorbents. Polymer electrospun fiber adsorbents with high adsorption capacity could be obtained by blend electrospinning, grafting modification, surface coating, incorporating functional material, imprinting, etc.

As a common synthetic polymer, polyacrylonitrile (PAN) fiber is an easily accessible and low-cost polymeric matrix with abundant cyano groups (C≡N). Raw PAN shows little adsorption capacity toward heavy metals and the adsorption capacity is dependent on a number of functional groups attached to the cyano groups [44, 45]. Many efforts have been focused on the grafted electrospun PAN fibers for the removal of heavy metals. Neghlani and coworkers prepared aminated polyacrylonitrile (APAN) nanofibers via the modification of electrospun PAN fibers by diethylenetriamine [46]. Their adsorption behaviors toward Cu^{2+} were examined and the aminated PAN nanofibers had a five times larger adsorption capacity than the aminated PAN microfibers, indicating the advantage of electrospun nanofibers. The maximum adsorption capacity based on the Langmuir model for Cu(II) ions was 116.52 mg/g. The thermodynamic parameters revealed that the adsorption process was endothermic and could be regarded as a chemical adsorption process. In addition to amine groups, some other functional groups are also grafted onto the surface of PAN fibers. Li et al. reported a thioamide-group chelating nanofiber membrane for the efficient adsorption of gold ions [47]. The chelating nanofiber membrane was prepared by the cross-linking and thioamide functionalization of electrospun PAN fibers. The preparation process and corresponding SEM images are shown in Fig. 5.4a–d. The adsorption processes toward Au(III) ions fit well with the pseudo-second-order rate equation and Langmuir adsorption isotherm. The maximum adsorption capacities toward Au(III) ions were 15.86, 23.50, and 34.60 mmol/g, when the temperature was 298, 323, and 348 K, respectively (Fig. 5.4e). In addition, the authors proposed a possible mechanism for the Au(III) adsorption by thioamide-group chelating nanofiber membranes. This adsorption process may involve a reduction process by reducing Au(III) to Au(I) and Au(0), then Au nanoparticles, Au(III) and Au(I) were adsorbed by thioamide-group chelating nanofibers (Fig. 5.4f). In another grafting work, Zhao and coworkers prepared a phosphorylated PAN-based nanofiber mat by the electrospinning technique followed by cross-linking, amination, and phosphorylation grafting modification with the content of phosphorous of 5.45 mg/g [48]. The grafted fibers showed good adsorption capacities toward Pb^{2+}, Cu^{2+}, and Ag^+ ions. All the adsorption kinetics followed the pseudo-second-order rate equation. Comparing with cross-linked and aminated fibers, the phosphorylated PAN nanofibers had higher adsorption equilibrium amounts toward Pb^{2+}, Cu^{2+}, and Ag^+ ions, suggesting the improvement of the adsorption capacity after phosphorylation. Moreover, the regeneration experiments revealed that the phosphorylated PAN nanofiber mats also had high recyclable removal efficiency.

Surface coating is another method to functionalize the PAN electrospun fibers. Pan's group reported conducting polymers-coated PAN electrospun fibers. They prepared polypyrrole (PPy) and polyaniline (PANi)-coated PAN core/shell nanofiber mats, respectively [49, 50]. The two composite fibers were used as adsorbents for Cr(VI) ions from aqueous solution. The adsorption processes by both the adsorbents followed a pseudo-second-order kinetics model and fit well with the Langmuir isotherm model. The PAN/PPy fibers had a maximum adsorption capacity of 61.80 mg/g at 25 °C and the value was 71.28 mg/g for PAN/PANi fibers. The positive values of $\Delta H°$ for both the adsorbents confirmed that the adsorption processes are endothermic. Recently, our group demonstrated the fabrication of the amino-rich hydrothermal carbon-coated electrospun polyacrylonitrile fiber (PAN@NC)

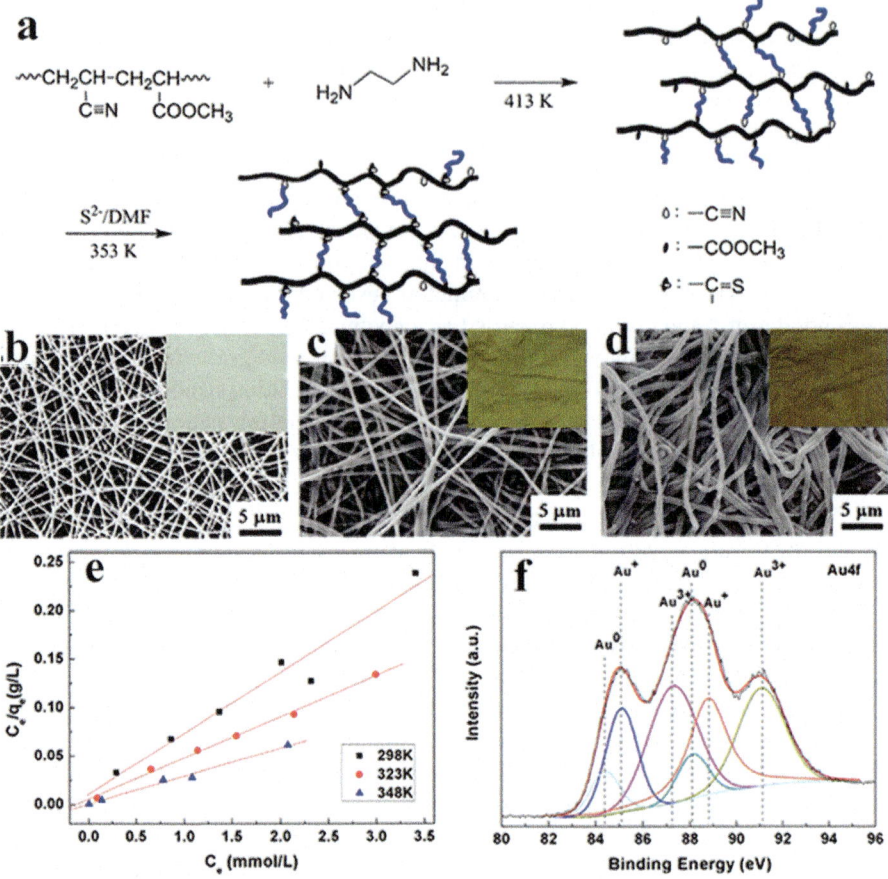

Fig. 5.4 (**a**) The synthetic scheme of thioamide-group chelating nanofibers based on PAN nanofibers. SEM and photo images of PAN nanofiber membranes (**b**), cross-linked nanofiber membranes (**c**), thioamide-group chelating nanofiber membranes (**d**). (**e**) Langmuir plot of Au(III) ions adsorption on the thioamide-group chelating nanofibers at 298, 323, and 348 K. (**f**) XPS spectra of Au_{4f} after adsorption of Au(III) ions. Adapted from Ref. [47] and reprinted with permission from Elsevier

adsorbents through one-step hydrothermal carbonization approach assisted by diethylenetriamine using electrospun PAN fibers as the templates (Fig. 5.5a) [51]. Traditional hydrothermal carbonaceous materials were in the form of powder which restricted the practical applications. The amino-rich hydrothermal carbon-coated electrospun polyacrylonitrile fiber had good flexibility and mechanical property. The SEM image is shown in Fig. 5.5b. The adsorption behavior toward Cr(VI) was studied. Based on the Langmuir isotherm model, the maximum adsorption capacity toward Cr(VI) was 290.70 mg/g (Fig. 5.5c). More importantly, the adsorbent could be recycled via a simple treatment and the removal efficiency still kept above 90% after five cycles. In addition to PAN, many other synthetic polymers, including

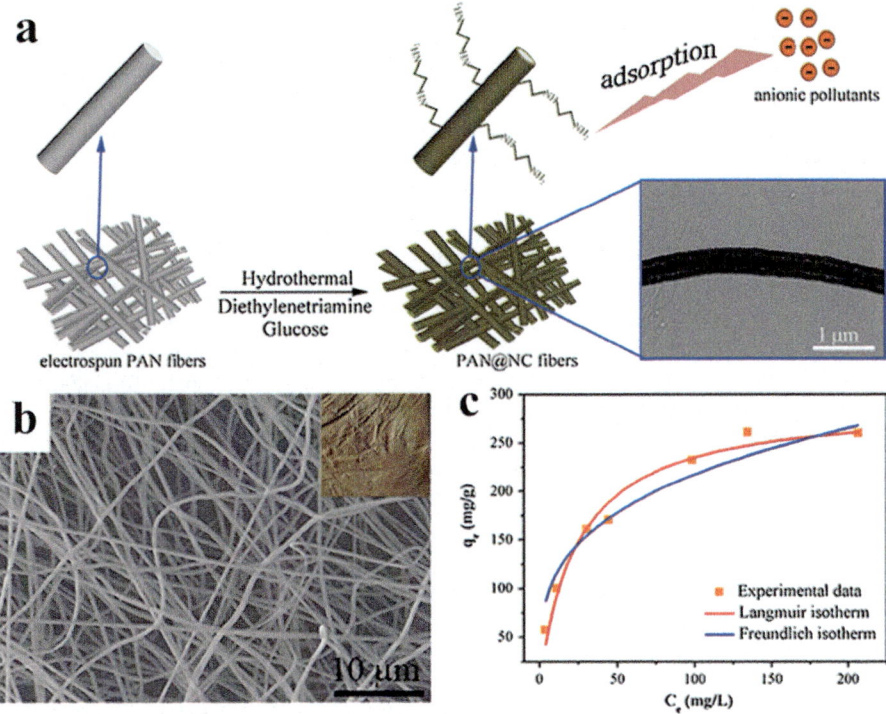

Fig. 5.5 (**a**) Fabrication procedure of amino-rich hydrothermal carbon-coated electrospun PAN (PAN@NC) fiber adsorbents. (**b**) SEM image of PAN@NC fibers. (**c**) Adsorption isotherms for Cr(VI). Adapted from Ref. [51] and reprinted with permission from Elsevier

polystyrene, polyindole, poly(vinyl alcohol), polyvinylchloride etc., were also electrospun to prepare organic electrospun nanofibers, which have been summarized in Table 5.2.

Natural polymers are low-cost and renewable; moreover, the natural polymer itself has many functional groups, such as amino group, carboxyl group, hydroxyl groups, etc.. Electrospun natural polymer nanofibers are widely explored as the heavy metal adsorbents. Cellulose is the most abundant natural polymer with desirable chemical and mechanical properties. However, the insolubility of cellulose in most of the solvents limits its electrospinning. So cellulose derivatives, which are readily soluble in volatile solvents, are used as the alternative of cellulose. Cellulose acetate (CA) is a useful material in industry, and it possesses high modulus, adequate flexibility, and tensile strength. Moreover, CA can also be transformed into nanofiber materials by electrospinning using appropriate solvent [78, 79]. Cellulose acetate electrospun fiber-based heavy metal adsorbents have received researchers' extensive attention. To obtain high adsorption ability, functionalization of the CA fibers is often conducted. Tian and coworkers prepared poly(methacrylic acid) (PMAA) modified cellulose acetate nonwoven membrane for heavy metal ion

Table 5.2 Organic electrospun nanofibers as heavy metal adsorbents

Adsorbent	Preparation	Heavy metal	Optimum pH	Adsorption capacity (mg/g)	Reference
3-aminopropyltriethoxysilane-modified PAN nanofibers	Electrospinning and modification	Th^{4+}	4	249.4	[52]
		U^{6+}	5	193.1	
		Cd^{2+}	6	69.5	
		Ni^{2+}	6	138.7	
Aminated electrospun PAN nanofiber	Electrospinning and grafting	Cu^{2+}	4	150.6	[53]
		Ag^+	4	155.5	
		Fe^{2+}	4	116.5	
		Pb^{2+}	4	60.6	
Polyethylenediaminetetraacetic acid-modified PAN nanofibers	Electrospinning and modification	Cd^{2+}	7	66.2	[54]
		CrO_4^{2-}	3	32.7	
Aminated PAN nanofibrous affinity membranes	Wet-electrospinning and amination	Pb^{2+}	5.5	1520.0	[55]
Thioamide-group chelating nanofiber membranes	Electrospinning and grafting	$AuCl_4^+$	3	3124.4	[47]
Phosphorylated PAN-based nanofiber	Electrospinning and grafting	Pb^{2+}	6	177.0	[48]
		Cu^{2+}	6	131.4	
		Ag^+	6	237.5	
		Cd^{2+}	6	37.3	
Acyl thioacetamide-group chelated nanofiber	Electrospinning and grafting	Ag^+	–	110.2	[56]
Polyacrylonitrile/polyaniline core/shell nanofiber	Electrospinning and in-situ polymerization	CrO_4^{2-}	2	71.3	[49]
Polyacrylonitrile/polypyrrole core/shell nanofiber	Electrospinning and in-situ polymerization	CrO_4^{2-}	2	61.8	[50]
Thiol functionalized cellulose-based electrospun nanofiber	Electrospinning and modification	Cu^{2+}	6	19.6	[57]
		Cd^{2+}	6	34.7	
		Pb^{2+}	6	31.0	
Cellulose–camphor soot nanofibers	Blending electrospinning	UO_2^{2+}	6	410.0	[58]
Oxolane-2,5-dione modified electrospun cellulose nanofibers	Electrospinning and modification	Cd^{2+}	5.8	325.9	[59]
		Pb^{2+}	5.8	207.2	
Acrylamide-plasma treated electrospun polystyrene nanofiber	Electrospinning and plasma treatment	Cd^{2+}	5	49.5	[60]
		Ni^{2+}	5	15.1	
Amidoxime functionalized polyindole nanofibers	Electrospinning and surface modification	Pb^{2+}	5	307.4	[61]
		Cd^{2+}	5	108.5	

(continued)

Table 5.2 (continued)

Adsorbent	Preparation	Heavy metal	Optimum pH	Adsorption capacity (mg/g)	Reference
Electrospun polyarylene ether nitrile nanofibrous mats	Electrospinning and chemical activation	Cu^{2+}	–	52.8	[62]
Electrospun chitosan/baker's yeast nanofibers	Blending electrospinning	UO_2^{2+}	6	219.0	[63]
		Th^{4+}	6	131.0	
Mesoporous chitosan/poly (ethylene oxide) nanofiber	Blending electrospinning	Cu^{2+},	–	120	[64]
		Zn^{2+}	–	117	
		Pb^{2+}	–	108	
Keratin/polyamide6 blend nanofibers	Blending electrospinning	Cu^{2+}	5.8	103.5	[65]
Electrospun nanofibrous polyethylenimine mat	Methacrylation and electrospinning	CrO_4^{2-}	3	108.9	[66]
		$HAsO_4^{2}$	3	233.6	
Nafion/polyvinyl alcohol nanofiber membrane	Blending electrospinning	Cu^{2+}	6	59.1	[67]
		Cr^{3+}	6	42.5	
		Co^{2+}	6	24.7	
		AsO_2^{+}	6	22.7	
Poly(ethyleneimine) nanofibrous affinity membrane	Wet-electrospinning	Cu^{2+}	5	67.2	[68]
		Cd^{2+}	5	116.9	
		Pb^{2+}	5	90.0	
Wool keratose/silk fibroin blend nanofibers	Electrospinning and post-treatment	Cu^{2+}	7	2.9	[69]
Electrospun zein nanoribbons	Coaxial electrospinning	Pb^{2+}	7	89.4	[70]
Capsular polypyrrole hollow nanofibers	Electrospinning, calcination, and vapor-phase polymerization	CrO_4^{2-}	2	839.3	[71]
Polydopamine-mediated surface functionalization of electrospun nanofibrous membranes	Electrospinning and surface modification	Cu^{2+}		33.6	[72]
Saccharomyces cerevisiae Loaded nanofibrous mats	Electrospinning and electrospraying	Pb^{2+}	6.5	238.0	[73]
Photo-cross-linked electrospun poly(vinyl alcohol)-based nanofiber	Electrospinning and cross-linking	Pd^{2+}	1.1	112.4	[74]
		$PtCl_6^{2-}$	1.1	69.9	
Bacteria-immobilized electrospun fibrous polymeric webs	Electrospinning and immobilization	CrO_4^{2-}	7	115.7	[75]

(continued)

Table 5.2 (continued)

Adsorbent	Preparation	Heavy metal	Optimum pH	Adsorption capacity (mg/g)	Reference
Pb(II) ion-imprinting electrospun cross-linked chitosan nanofiber mats	Electrospinning and imprinting	Pb^{2+}	7	110.0	[76]
Thermally cross-linkable chitosan-based nanofibrous mats	Electrospinning and cross-linking	Cu^{2+}	7	79.0	[77]

adsorption by electrospinning and surface modification [80]. The surface of electrospun CA fibers was modified by grafting PMAA using Ce^{4+} initiated polymerization, which provided adsorptive –COOH groups. The adsorption and desorption of Cu^{2+}, Hg^{2+}, and Cd^{2+} onto the PMAA-modified CA fibers were investigated. The effect of pH values, initial ion concentrations, and reuse were discussed. Higher initial pH value corresponded to higher adsorption capacity, and the modified CA fibers had high adsorption selectivity toward Hg^{2+}. This study provided some information for fabricating low-cost, highly efficient adsorbents for heavy metal ion removal. Stephen et al. reported oxolane-2,5-dione modified cellulose nanofibers [59]. First, the cellulose nanofibers were obtained by the deacetylation of electrospun cellulose acetate fibers using NaOH solution (0.3 M). Then the cellulose nanofibers were functionalized by oxolane-2,5-dione (Fig. 5.6a and b). Cd^{2+} and Pb^{2+} were selected to study the adsorption property of oxolane-2,5-dione modified cellulose nanofibers. Due to the increased surface area and pore volume of cellulose nanofibers, the functionalized cellulose nanofibers showed improved adsorption property compared to the functionalized cellulose bulk fibers. The adsorption isotherm toward Cd^{2+} and Pb^{2+} by oxolane-2,5-dione modified cellulose nanofibers is shown in Fig. 5.6c. The adsorption of both Cd^{2+} and Pb^{2+} was better described by the Freundlich isotherm model. Regenerability experiments indicated that the adsorption capacity did not change significantly after three times.

Chitosan is another natural polymer derived from the chitin, which is often used for metal ion removal due to the large number of amino and hydroxyl functional groups on chitosan. Considering the electrospinnability and mechanical property, synthetic polymers such as poly(ethylene oxide), poly(vinyl alcohol), poly(lactic acid), poly(caprolactone), and nylon-6 are often blended with chitosan to improve the properties of chitosan nanofibers. Shariful and coworkers prepared poly(ethylene oxide) (PEO)/chitosan composite nanofiber membrane by the electrospinning technique for adsorption of nickel (Ni), cadmium (Cd), lead (Pb), and copper (Cu) from aqueous solution [64]. The average diameter and surface area of the composite nanofiber membrane were 98 nm and 312.2 m^2/g, respectively. The adsorption of lead, copper, cadmium, and nickel onto the membrane was in the following order: $Pb^{2+} < Cd^{2+} < Cu^{2+} < Ni^{2+}$ in the adsorption selectivity study. Both the pseudo-first-order and pseudo-second-order kinetic models could describe the adsorption processes. The equilibrium data of nickel, cadmium, copper, and lead adsorption fitted

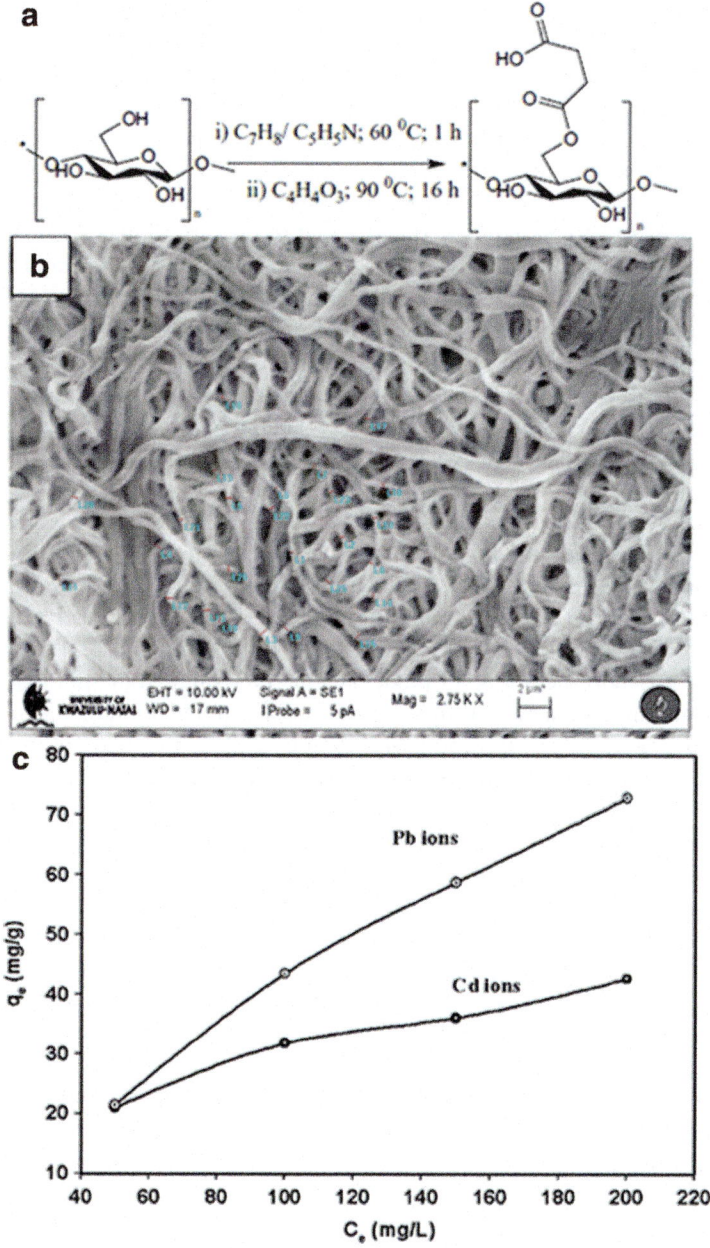

Fig. 5.6 (**a**) Reaction scheme for the functionalization of cellulose to cellulose-g-oxolane-2,5-dione. (**b**) SEM micrograph of cellulose-g-oxolane-2,5-dione nanofibers. (**c**) Adsorption isotherm for Pb and Cd. Adapted from Ref. [59] and reprinted with permission from Elsevier

better with the Langmuir model. The obtained thermodynamic parameters indicated that the adsorption processes were feasible, spontaneous, and endothermic.

To improve the adsorptive selectivity and capacity, Tong et al. synthesized novel Pb^{2+} ion-imprinting electrospun cross-linked chitosan nanofiber mats by one-step electrospinning and ion-imprinting method [81]. The in-situ ion-imprinting method could prevent the amino group from excessively cross-linking during the cross-linking process and the adsorption capacity had an improvement. The adsorptive selectivity was due to recognition sites formed in the cross-linked polymer matrix by imprinted molecules or ions. The Pb^{2+} ion-imprinting electrospun chitosan nanofibers showed excellent selectivity for the target ion Pb^{2+} with the equilibrium adsorption capacity in the order of $Pb^{2+} > Cu^{2+} > Zn^{2+} > Cd^{2+} > Ni^{2+}$. The adsorption of Pb^{2+} ions followed the pseudo-second-order model and Langmuir model. Moreover, the imprinted chitosan nanofiber mats could be regenerated and reused again. These relevant results indicated it as a potential material for application in the field of wastewater treatment. A variety of other natural polymers have also been electrospun into nanofibers successfully for the removal of heavy metal ions. For example, zein nanoribbons were reported by a modified coaxial electrospinning process [70]. To exploit the favorable interactions between metal and protein, zein nanoribbons were utilized for the application in the removal of Pb^{2+} from polluted water. The adsorption equilibrium could be obtained in 60 min for Pb^{2+} solutions with initial concentrations of 100, 150, and 200 mg/L. The adsorption processes followed the pseudo-second-order model and the Langmuir isotherm model. On the basis of isotherm parameters from Langmuir model, the maximum adsorption was 89.37 mg/g. After five absorption/desorption cycles, the adsorption capacity remained up to 82.3%. All the results demonstrated the potential applications of these zein nanoribbons in the treatment of polluted water. In addition to this, electrospun natural polymers, such as keratin and silk fibroin, were also studied as the metal adsorbents. The typical examples are shown in Table 5.2.

Other peculiar compositions and architectures are also reported to enrich the adsorbents. Zhao et al. fabricated capsular polypyrrole hollow nanofibers (PPy-HNFs) via in-situ vapor-phase polymerization approach using electrospun V_2O_5 fiber template [71]. The morphology was characterized by SEM and TEM images. The adsorption of Cr(VI) onto PPy-HNFs was studied. The adsorption process could be described using the pseudo-second-order model and the adsorption capacity was up to 839.3 mg/g for 200 ppm at pH = 2. After five adsorption/desorption cycles, the adsorption capacity still maintained 75%. The results demonstrated the possibilities of using this capsular PPy-HNF membrane for heavy metal removal from aqueous solution.

Another interesting study was reported by Uyar group [75]. They described an isolated *Morganella morganii STB5* strain on electrospun PS and PSU fibrous webs. *Morganella morganii STB5* is a bacterial strain previously isolated which could be used for the effective removal of Cr(VI) under neutral conditions. The immobilization process could keep the cell viability and biochemical activity. The bacteria-immobilized electrospun PS and PSU fibrous webs showed removal yields of 93.60 and 93.79% for 10 mg/L, 99.47 and 90.78% for 15 mg/L, and 70.41 and 68.27% for

25 mg/L of initial hexavalent chromium within 72 h, respectively. The two webs could be reused several cycles and could be stored for 15 days without exhibiting losses in attached cell numbers or Cr(VI) remediation capabilities. Table 5.2 shows a summary of a variety of inorganic electrospun nanofibers as heavy metal adsorbents.

5.2.3   Organic/Inorganic Composite Nanofibers

Each type of the above described adsorbent has unique features. Inorganic electrospun nanofibers have high specific surface area and small size; however, the powder form makes them difficult to be separated and used in practical applications. Organic composite nanofibers have good membrane morphology with satisfying mechanical properties; however, most organic nanofibers themselves show low binding capacity toward heavy metals, which need to be modified further. For these reasons, organic/inorganic composite nanofibers have attracted much interest as potential heavy metal adsorbents. Organic/inorganic composite nanofibers could combine the advantages of both the components and have the synergistic effect for the adsorption process. In-situ blend electrospinning and post-treatment are main methods to prepare organic/inorganic composite nanofibers. Various types and compositions have been reported, such as synthetic polymer/inorganic matter and natural polymer/inorganic matter composite nanofibers.

As common inorganic materials, nano-zeolites have high efficiency for heavy metals treatment due to their unique properties, such as high surface area and more accessible active sites. Rad et al. reported the polyvinyl alcohol (PVA)/NaX nano-zeolite nanocomposite nanofibers from PVA/zeolite solution [82]. The surface area (SBET) of synthesized PVA/NaX nanofibers was found to be 212 m^2/g. The composite nanofibers were investigated for the removal of Ni^{2+} and Cd^{2+} ions from aqueous solutions. The adsorption capacity toward Cd^{2+} ions was higher than that of Ni^{2+} ions. Both the adsorption processes followed the pseudo-second-order kinetic model and fitted well with the Langmuir isotherm model. The thermodynamic parameters indicated that the adsorption processes were endothermic and spontaneous at studied conditions. In addition, the nanofibers could be reused repeatedly. To further functionalize the adsorbents, Ang's group prepared a three-component electrospun adsorbent of the chitosan/PVA/zeolite composite fiber membrane [83]. The resulting membrane was stable in distilled water, acidic, and basic media for 20 days. The adsorption performance toward Cr(VI), Fe(III), and Ni(II) ions was studied. Langmuir isotherm and the pseudo-second-order kinetic models could well describe the adsorption processes. Furthermore, the adsorption capacity of the nanofiber changed little after five runs. Graphene oxide (GO), as a new carbon-based nanomaterial, has attracted increasing attention as adsorbent materials. Tan and coworkers fabricated PVA/GO nanofibers by the electrospinning of PVA/GO solution [84]. Cu^{2+} and Cd^{2+} were adsorbed as the target ions to investigate the adsorption performances of the PVA/GO composite nanofibers. The adsorption followed the pseudo-second-order kinetic model and the Langmuir model. Notably, the adsorption could reach equilibrium within 25 min. As the FTIR and XPS results show, carboxyl and carbonyl groups on the surface of the adsorbent contributed to the adsorption of Cu^{2+} and Cd^{2+}. These obtained results suggested that PVA/GO composite nanofibers could be efficient, nontoxic, recoverable, and economical adsorbents in heavy metals removal applications. PVA showed almost no binding capacity toward heavy metals and it acted as the flexible carrier. Najafabadi et al. developed a novel electrospun chitosan/GO nanofibrous adsorbent to combine the adsorption capacities of chitosan and GO [85]. The adsorption behaviors of Cu^{2+}, Pb^{2+}, and Cr(VI) metal ions from aqueous solutions were investigated. The adsorption processes followed double-exponential mechanism and the Redlich–Peterson isotherm model. The maximum adsorption capacity of Pb^{2+}, Cu^{2+}, and Cr(VI) onto chitosan/GO composite nanofibers based on Langmuir isotherm were 461.3, 423.8, and 310.4 mg/g at 45 °C, and the obtained thermodynamic parameters showed their endothermic and spontaneous nature. Moreover, the chitosan/GO nanofibers could be reused frequently without almost any significant loss in adsorption capacity. Other inorganic materials, such as hydroxyapatite, boehmite, metal-oxide, etc., are also applied to prepare the organic/inorganic composite electrospun nanofiber adsorbents (Table 5.3).

In addition to the blend electrospinning, post-treatment is another method to obtain these composite nanofibers. The organic/inorganic composite electrospun nanofibers obtained by this method often have hierarchically structures, which could increase the surface area and thus enhance the adsorption capacity. Xu et al. fabricated "flower-Like" polyamide 6 (PA6)@Mg(OH)$_2$ composite fibers (Fig. 5.7a and b) by electrospinning combined with hydrothermal strategy [107]. This novel morphology could also increase the specific surface area of electrospun fibers. The adsorption of Cr(VI) onto the PA6@Mg(OH)$_2$ composite fibers was studied. The composite fibers showed good performance for the removal of Cr(VI) and the adsorption capacity could reach 296.4 mg/g when the concentration of Cr(VI) was 110 mg/g (Fig. 5.7c). Meanwhile, the fiber membrane could be easily separated from Cr(VI) solutions and exhibited excellent good recyclable performance (Fig. 5.7d). Sun and coworkers prepared hierarchical aminated polyacrylonitrile (PAN)/γ-AlOOH electrospun composite nanofibers by a combination of electrospinning process, chemical modification, and hydrothermal reaction [96]. The adsorption behaviors toward Pb^{2+}, Cu^{2+}, and Cd^{2+} ions in aqueous solution were investigated. All the adsorption processes followed the pseudo-second-order rate equation and fitted well with the Langmuir isotherm model. The maximum monolayer adsorption capacities for Pb(II), Cu^{2+}, and Cd^{2+} were 180.83 mg/g, 48.68 mg/g, and 114.94 mg/g, respectively. Adsorption mechanism analysis proved that the amine groups and γ-AlOOH crystals synergistically contributed to the high adsorption activity of the hierarchical aminated PAN/γ-AlOOH composite nanofibers. The obtained results suggested their potential application in heavy metal ions removal from wastewater effluents. Ji's group also synthesized the brush-like hydroxyapatite (Hap)/PAN composite fibers through electrospinning and hydrothermal treatment [99]. Adsorption experiments for Pb^{2+} removal by the PAN/Hap composite fiber mat were conducted. Based on the Langmuir isotherm model, the maximum adsorption

Table 5.3 Organic/inorganic composite electrospun nanofibers as heavy metal adsorbents

Adsorbent	Preparation	Heavy metal	Optimum pH	Adsorption capacity (mg/g)	Reference
Poly(vinyl alcohol)/grapheme oxide nanofibers	Blending electrospinning	Cu^{2+}	5.8	25.8	[84]
		Cd^{2+}	5.8	32.4	
Electrospun PVA/zeolite nanofibers	Blending electrospinning	Cd^{2+}	–	838.7	[82]
		Ni^{2+}	–	342.8	
PVA/tetraethylorthosilicate/aminopropyl triethoxysilane composite nanofiber	Sol–gel/electrospinning method	Cd^{2+}	6	327.3	[86]
A boehmite nanoparticle impregnated electrospun fiber membrane	Blending electrospinning	Cd^{2+}	7	0.2	[87]
Chitosan/graphene oxide composite nanofibers	Blending electrospinning	Pb^{2+}	6	461.3	[85]
		Cu^{2+}	6	423.8	
		CrO_4^{2-}	3	310.4	
Chitosan/hydroxyapatite composite nanofiber	Blending electrospinning	Pb^{2+}	–	296.7	[88]
		Ni^{2+}	–	213.8	
		Co^{2+}	–	180.2	
Chitosan/cobalt ferrite nanofibrous adsorbent	Blending electrospinning	Pb^{2+}	6	283.3	[89]
		CrO_4^{2-}	3	179.1	
Chitosan/TiO_2 composite nanofibrous adsorbents	Blending electrospinning	Cu^{2+}	6	710.3	[90]
		Pb^{2+}	6	579.1	
Chitosan/(polyvinyl alcohol)/zeolite electrospun composite nanofibrous membrane	Blending electrospinning	CrO_4^{2-}	–	8.84	[83]
		Fe^{3+}	–	6.16	
		Ni^{2+}	–	1.77	
Electrospun PVA/ZnO nanofiber adsorbent	Blending electrospinning	UO_2^{2+}	5	370.8	[91]
		Cu^{2+}	5	157.4	
		Ni^{2+}	5	94.4	
Fe^{3+} ion immobilized poly(vinyl alcohol) nanofibers	Blending electrospinning	AsO_2^-	7	67	[92]
		$HAsO_4^{2-}$	7	36	

(continued)

Table 5.3 (continued)

Adsorbent	Preparation	Heavy metal	Optimum pH	Adsorption capacity (mg/g)	Reference
NH2-functionalized cellulose acetate (CA)/silica composite nanofibrous membranes	Hydrolysis and electrospinning	CrO_4^{2-}	1	19.5	[93]
Electrospun cellulose nanofibers modified by montmorillonite	Blending electrospinning	CrO_4^{2-}	3	–	[94]
Zero-valent iron nanoparticle-immobilized electrospun nanofibers	Electrospinning and reduction	Cu^{2+}	4.5–5.0	107.8	[95]
Hierarchical aminated PAN/ γ–AlOOH electrospun composite nanofibers	Electrospinning process, chemical modification, and hydrothermal reaction	Pb^{2+} Cu^{2+} Cd^{2+}	5 6 6	233.10 43.90 144.30	[96]
Hierarchically nanofibrous membrane containing of thermal plastic elastomer ester nanofibers core and iron oxides shell	Electrospinning and hydrothermal method	CrO_4^{2-}	2	–	[97]
Thermal plastic elastomer ester@AlOOH nanofibers	Electrospinning and hydrothermal method	CrO_4^{2-}	2	16.66	[98]
Brush-like polyacrylonitrile–hydroxyapatite composite fibers	Electrospinning and hydrothermal method	Pb^{2+}	5.2	433	[99]
Hierarchical nanofiber mat (polyacrylonitrile/polypyrrole/manganese dioxide)	Electrospinning and post-treatment	Pb^{2+}	6	172.41	[100]
Aminated Fe3O4@SiO2/PVA nanofiber adsorbent	Modification and electrospinning	UO_2^{2+}	4.5	68.9	[101]
Amino functionalized mesoporous polyvinyl pyrrolidone/SiO2 composite nanofiber	Hydrolysis and electrospinning	Cr^{3+}	7.0	97	[102]
A novel PVP/CeO2/TMPTMS nanofiber adsorbent	Blending electrospinning	Pb^{2+} Cu^{2+}	6 6	272.3 263.4	[103]
A novel PAN–TiO2 nanofiber adsorbent modified with aminopropyltriethoxysilane	Blending electrospinning	Th^{4+} Ni^{2+} Fe^{2+}	5 6 5	250 147 80	[104]
Phosphine-functionalized electrospun poly(vinyl alcohol)/silicananofibers	Electrospinning and chemical treatment	Mn^{2+} Ni^{2+}	6 6	228.60 224.90	[105]
Thiol-functionalized mesoporous PVA/SiO2 composite nanofiber	Sol–gel process and electrospinning	Cu^{2+}	5–6	489.12	[106]

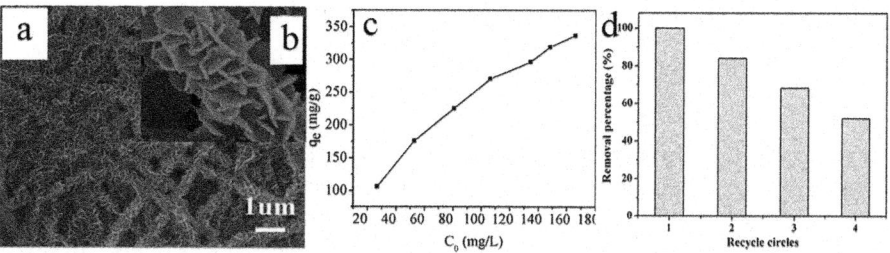

Fig. 5.7 (**a, b**) SEM images of PA6@Mg(OH)$_2$ composite fibers. (**c**) Adsorption isotherm of Cr(VI). (**d**) The recyclable properties of PA6@Mg(OH)$_2$ composite nanofibrous membrane for Cr(VI) removal. Adapted from Ref. [107] and reprinted with permission from Elsevier

capacity for Pb^{2+} was 433 mg/g. The adsorption process could be explained by a two-step mechanism: rapid surface complexation followed by partial dissolution of hydroxyapatite and precipitation of pyromorphite. The experiment results indicated that the PAN/Hap composite fiber mat could be a promising adsorbent for heavy metal ions from industrial wastewater. Zero-valent iron nanoparticles (ZVI NPs) have also received much interest in environmental applications. Xiao et al. fabricated ZVI NPs-immobilized hybrid electrospun polymer nanofibrous mats [95]. The hybrid nanofibrous mats were obtained by electrospinning polyacrylic acid (PAA)/polyvinyl alcohol (PVA)/multiwalled carbon nanotubes (MWCNTs) mixture polymer solution. Then, ZVI NPs were further immobilized onto the nanofibrous mats. The ZVI NPs-immobilized hybrid nanofibrous mats were used for removing Cu(II) ions. The influence of factors such as pH, contact time, solution concentration, and solution ionic strength was examined. On the basis of Langmuir model, the maximum adsorption capacity after the immobilization of ZVI NPs increased to 107.8 from 66.8 mg/g. The adsorption of Cu^{2+} occurred via chemical reduction and deposition on the ZVI NP surfaces to form Fe/Cu alloy. The study suggests that the ZVI NPs-immobilized hybrid nanofibrous mat could be a useful material for the removal of Cu^{2+} ions.

Another interesting work was reported by Li and Zhangs' group. They fabricated a series of mesoporous SiO$_2$-based composite nanofiber membranes [102, 106, 108, 109]. First, PVA/SiO$_2$ gels were prepared and then were electrospun. The flexible SiO$_2$-based composite nanofiber membranes with mesostructure were obtained without calcination. PVA acted as the flexible carrier and made the composite nanofibers have good film-forming properties. The mesoporous SiO$_2$-based composite nanofiber membranes were used for adsorption of heavy metal ions from aqueous solution. To further improve the adsorption capacity, thiol, thioether, amino, and phosphine groups were also introduced into these mesoporous SiO$_2$-based composite nanofibers. The typical examples are shown in Table 5.3. Organic/inorganic composite nanofibers have the synergistic effect between the organic component and inorganic component for adsorption, making them become the research hotspot and tendency.

5.3   Electrospun Filters for Heavy Metals Removal

Filtration processes for water purification treatment are of great interest for practical applications due to the consumption of money and time, space saving, and cost-effectiveness. Membrane filtration technologies with different types of membranes have shown great potential for heavy metal removal [110–112]. Electrospun nanofiber membranes have high tunable porosities, interconnected pore structures, and relatively high mechanical strength, making them suitable to act as filters. However, the pore size among the electrospun fibers is much larger than the ion radius of heavy metal ions. Filtration processes for heavy metal ions removal by the electrospun filters are mainly dependent on the binding interaction between the filters and the metal ions. Most of the researches focus on the static adsorption performance, which serves as the foundation of the filtration process. Thus, the development of electrospun filters for heavy metals is limited. Below are some studies which use electrospun fiber membranes for the dynamic filtration of heavy metals.

Mechanical strength is an important factor for filtration applications. In most cases, the electrospun filters are made of functionalized organic electrospun nanofibers. Natural polymers are often used to prepare low-cost electrospun filters. For example, Denkbaş and coworkers prepared silk fibroin/nylon-6 nanofiber matrices by electrospinning [113]. The fiber surface was functionalized with calcium phosphate crystals to increase the affinity of divalent heavy metals. The optimum Cu^{2+} removal was obtained at the condition of pH: 5.0, seven layers of nanofiber, 30 mg/l of initial Cu^{2+} concentration, and 0.1 mL/min of flow rate. 32% of Cu^{2+} ions could be removed by the continuous flow system. Another silk fibroin-based electrospun filter was reported by Cui's group [114]. The silk fibroin/cellulose acetate composite nanofibrous membranes were prepared by electrospinning and followed by ethanol treatment. For the filtration process, the solution containing 100 ppm Cu^{2+} was pumped through the nanofiber membranes using a peristaltic pump with a volumetric flow rate of 2.0 mL/min. The composite fiber membranes with 20% content of cellulose acetate had a good performance for the removal of Cu^{2+}, and the maximum milligrams per gram of Cu^{2+} removed was 22.8 mg/g. Moreover, the silk fibroin/cellulose acetate blend nanofiber membranes exhibited a higher Cu^{2+} removal capacity than that of pure silk fibroin or cellulose acetate nanofiber membranes. These studies provided the relatively comprehensive data for the silk fibroin-based electrospun filters for the removal of heavy metal ion in wastewater.

Some other natural polymers have also been electrospun to act as the filters. Randomly oriented wool-derived keratin nanofibers with a mean diameter of 240 nm were prepared by the electrospinning process [115]. The Cu^{2+} metal ion removal from aqueous solutions was by pumping the ion solution through the fiber membranes with a peristaltic pump. The experimental data had a good fit with both Langmuir and Freundlich isotherm models. The maximum removal capacity obtained from the Langmuir model was about 11 mg/g. By the Dubinin–Radushkevich model, the removal process of Cu^{2+} ions on keratin nanofiber membranes was through the ion exchange reactions. The competitive tests indicated that the selectivity order by keratin nanofiber membranes toward the considered metal ions was $Cu^{2+} > Ni^{2+} > Co^{2+}$. Chitosan is a widely used adsorption material and is suitable for the preparation of filters. Lee et al. developed biodegradable nanofibrous polylactide/chitosan membranes for the filtration of silver ions [116]. Both polylactide and chitosan were popular "green" eco-friendly materials. The polylactide/chitosan fiber membranes for the filtration test had a diameter of 40 mm and a thickness of 0.15 mm. The Ag^+ removal capacity by the nanofibrous polylactide/chitosan filter increased with the initial silver ion concentrations and the pH value, while it decreased with the temperature and the filtering rate of the solutions. Besides, this electrospun filter could be reused after the recovery process. Kit's group also fabricated the nanofibrous chitosan nonwoven filter [117]. The novel filter was prepared by electrospinning of chitosan/PEO blend solutions onto a spunbonded nonwoven polypropylene substrate. The filtration of Cr(VI) ions was conducted. The dynamic Cr(VI) binding capacity was up to 35 mg Cr(VI)/g chitosan, which showed promise for commercial application. Moreover, this chitosan-based nanofibrous filter had good antimicrobial properties along with a 2–3 log reduction in *Escherichia coli* bacteria cfu. The chitosan-based nanofibrous filter media offered immense potential application for filtration purposes.

Compared with natural polymers, synthetic polymers are more applicable as a filter due to their good strength. As previously stated, the binding capacity toward metal ions by electrospun fibers from traditional synthetic polymers is limited. In most cases, the electrospun nanofibers should be functionalized to improve the performance of filtration. Liu and coworkers reported a polyvinylamine grafted polyacrylonitrile electrospun nanofibrous microfiltration membrane, which was capable of removing Cr(VI) from contaminated water [118]. This filter membrane had a mean pore size of 0.40 ± 0.01 µm with a pure water flux of 645 ± 37 L/m²·h·psi. At a flow rate of 70 L/m²·h, the removal ability was 36.3 ± 2.6 mg Cr(VI) per gram of the membrane. In addition, the filter membrane could be regenerated using 0.1 mol/L of NaOH aqueous solution with the desorption ratio > 98%. Wu et al. developed polypyrrole (PPy)-embedded electrospun polyethersulfone nanofibrous membranes for the removal of silver ions [119]. The average diameter of obtained nanofibers ranged from 410 nm to 540 nm. The as-spun filter membrane had a diameter of 43 mm and a thickness of 1.3 mm for the tests. The maximum removal capacity of Ag (I) ions onto the polypyrrole/polyethersulfone nanofibrous membrane was 35.7 (Ag/PPy) (mg/g). The removal capacity increased with the initial metal ion concentrations and the pH value but decreased with the temperature and the filtering rate of the solutions. The results suggested that electrospun polypyrrole/polyethersulfone nanofibrous membranes could act as a potential filter for metal ion removal. Recently, our group reported a novel branched polyethylenimine (b-PEI) grafted electrospun polyacrylonitrile (PAN) fiber membrane for the Cr(VI) remediation [120]. The b-PEI grafted electrospun PAN fiber membrane was synthesized by electrospinning and a facile refluxing approach (Fig. 5.8a). The electrospun fibers showed good membrane-forming property, guaranteeing their application in the filtration purification process. The obtained b-PEI grafted electrospun PAN fibers possessed an excellent adsorption capacity toward Cr(VI) ($q_m = 637.46$ mg/g), which was higher than many

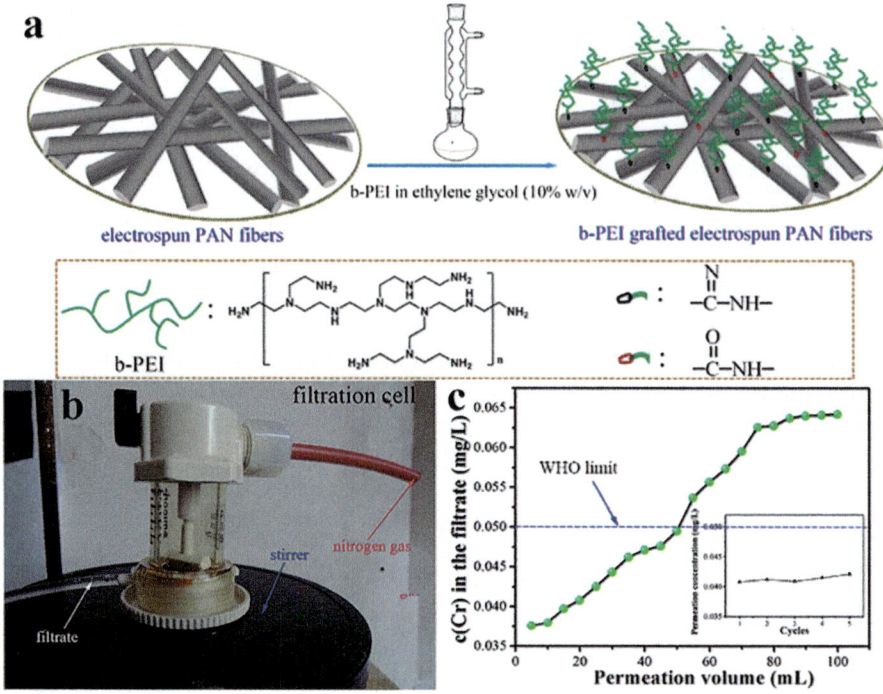

Fig. 5.8 (**a**) Schematic diagram of the synthesis of b-PEI grafted electrospun PAN fibers. (**b**) The filtration cell. (**c**) Breakthrough curves for Cr(VI) solutions through the bPEI-EPAN membrane (thickness: 150 μm) at pressure of 0.1 bar (the inset is regeneration-filtration cycle). Adapted from Ref. [120] and reprinted with permission from the Royal Society of Chemistry

of other adsorbents. In addition, b-PEI grafted electrospun PAN fibers could make the Cr(VI) concentration a conspicuous decrease from 10 or 5 mg/L to below 0.05 mg/L, which is recommended as the standard of drinking water by WHO. Based on the excellent adsorption performance, the dynamic filtration experiments were conducted (Fig. 5.8b). For the filtration experiments, the initial concentration of Cr(VI) solution is fixed at 5 mg/L and applied pressure is fixed at 0.1 bar. The results revealed that the b-PEI grafted electrospun PAN fiber filter could decrease the residual Cr(VI) in the filtrate below 0.05 mg/L from over 50 mL of solution at a high flux of 731.7 L·m^{-2}·h^{-1} (Fig. 5.8c). The data suggested that the b-PEI grafted electrospun PAN fiber membrane was a promising candidate for Cr(VI) remediation from wastewater.

Some other methods and filter structures are also studied to enrich the preparation of the electrospun filters. Li's group prepared anew thin film nanofibrous composite membranes through layer-by-layer assembly of chitosan/alginate on the surface of the electrospinning nanofibrous substrates for Cr(VI) removal [121]. The higher Cr(VI) removal efficiency was obtained when the layer-by-layer cycles increased. But the water permeation rate decreased with the increase of the layer-by-layer cycles. When the layer-by-layer cycle was 20.5, the total Cr(VI) removal efficiency was approximately 70% at the Cr(VI) concentration of 40 ppm. Compared with traditional membranes for heavy metal removal, these new thin film nanofibrous composite membranes exhibited simple fabrication process and low energy consumption. This method served as a novel strategy fabrication of similar nanofibrous composite membranes for removing other metal ions. Multilayer membrane composite is also an effective route to prepare filter membrane. Wang and coworkers developed a novel class of microfiltration membrane, based on a two-layered nanoscale polyacrylonitrile (PAN) fibers/microscale polyethylene terephthalate (PET) fibrous support infused with ultra-fine functional cellulose nanofibers (about diameter of 5 nm) [122]. The total thickness of microfiltration membrane was 140 ± 10 μm, and the pure water permeation rate was 1300 L/m^2 h/ psi. The removal capabilities for Cr(VI) and Pb^{2+} were 100 and 260 mg/g, respectively.

5.4   Conclusions and Perspective

In this chapter, we have described electrospun adsorbents and filters for the removal of heavy metals. Increasingly exciting studies have been performed due to the interesting characteristics of electrospun fibers such as high porosity, large surface-to-volume ratio, easy preparation, good membrane formation and flexibility. The filtration of heavy metals by the electrospun filters is mainly due to the adsorption process between the electrospun fibers and the heavy metals. Numerous papers have focused on the adsorption processes as the basis for the filtration. There are three main types of electrospun adsorbents involved in heavy metal treatment: inorganic nanofibers, organic nanofibers, and organic/inorganic composite nanofibers. Different types of fiber adsorbents have their individual advantages. Inorganic nanofibers have small fiber diameters and high surface-to-volume ratio, but they are brittle. Organic nanofibers have satisfying strength, but the raw electrospun organic nanofibers show low binding capacity toward heavy metals and the organic fibers should be modified to improve the adsorption capacity. Organic/inorganic composite nanofibers could combine the advantages of both the components and have the synergistic effect for the adsorption process. All these types of electrospun fibers have been proved to be capable of removing heavy metals based on the adsorption processes. Due to the good membrane formation and binding capacity, electrospun fiber membranes have been used as filters for the removal of heavy metals. In consideration of strength, organic nanofibers or organic/inorganic composite nanofibers are often studied for the filtration of metal ions. All the results show that the electrospun fibers have good potential in heavy metal wastewater treatment.

Although electrospun materials show great potential as heavy metal adsorbents or filters in wastewater treatment, many technical issues and challenges remain to be solved or addressed:

(1) Mechanical strength. There is a need for fabrication of high-mechanical-strength nanofiber filters with resistance to water flow in practical application. Due to the fragile characteristic, electrospun fibers are limited in the filtration field. Thus, it is still a challenge to fabricate electrospun nanofibers with high mechanical strength.

(2) Removal mechanism. Most of the studies focus on the design and preparation of various novel electrospun adsorbent or filter materials. Less attention is given to in-depth investigations of the heavy metals removal process and the understanding of the fundamental mechanisms and theories during this process. The mechanism including physical interaction and chemical interaction for heavy metals removal by electrospun adsorbents or filters is a scientific guidance for the design and preparation of electrospun materials. It is needful to give mechanistic insights into the heavy metal removal.

(3) Low-cost and reusability. For industrial application, the cost of the filters is an important consideration. Developing electrospun adsorbents or filters with low-cost and easy preparation is vital for commercialization. The working life is another important consideration in industrial applications. The adsorbents or filters which can be recycled many times are ideal materials to prolong the working life.

(4) Synergistic effect. Multicomponent electrospun fibers, such as organic/inorganic, are the research tendency for the preparation of satisfying adsorbents or filters by synergy with the multicomponents. Through the synergistic effect, researchers should develop the adsorbents or filters which can remove a variety of heavy metal ions at the same time and also show high mechanical strength.

(5) Multifunction. Most of the researchers explore the removal of one heavy metal. However, more than one heavy metal or contaminant exist in the wastewater. Thus, the development of adsorbents or filters which could remove multiple heavy metals or contaminants simultaneously is meaningful. Moreover, the adsorbents or filters with other functions, such as antibacterial property, catalytic property, etc., are needed to be investigated to develop multifunctional materials.

In conclusion, electrospun materials have been playing an increasingly important role in heavy metal adsorbent or filter fields. Most of the researchers focus on the development of novel materials as adsorbents or filters, and rarely considerations aim at the practical applications. Electrospun adsorbents or filters should be studied with more practical considerations, such as the removal efficiency, usability (mechanical property, ability against the interference, no secondary-pollution and stability), and reusability (working life and easy regeneration). Although there is a long way to go before they can be used in large-scale industrial applications, we believe that the above challenges may be solved and electrospun adsorbents or filters for heavy metals removal will find their way toward practical applications in the near future. This situation could provide more opportunities for the colleagues of the electrospinning research communities around the world.

References

1. Mahmud HNME, Huq AKO, binti Yahya R (2016) RSC Adv 6:14778–14791
2. Lin Y, Gritsenko D, Feng S, Teh YC, Lu X, Xu J (2016) Biosens Bioelectron 83:256–266
3. Rezania S, Taib SM, Din MFM, Dahalan FA, Kamyab H (2016) J Hazard Mater 318:587–599
4. Wang J, Shen H, Hu X, Li Y, Li Z, Xu J, Song X, Zeng H, Yuan Q (2015) Adv Sci 3:1500289
5. Alcamo J, Henrichs T, Rosch T (2000) Kassel world water series report no. 2, Centre for Environmental Systems Research, University of Kassel, Germany
6. Zou Y, Wang X, Khan A, Wang P, Liu Y, Alsaedi A, Hayat T, Wang X (2016) Environ Sci Technol 50:7290–7304
7. Fu F, Wang Q (2011) J Environ Manag 92:407–418
8. Kurniawan TA, Lo WH, Chan GY (2006) J Hazard Mater 129:80–100
9. Khin MM, Nair AS, Babu VJ, Murugan R, Ramakrishna S (2012) Energy Environ Sci 5:8075–8109
10. Srivastava NK, Majumder CB (2008) J Hazard Mater 151:1–8
11. Ngah WW, Hanafiah MAKM (2008) Bioresour Technol 99:3935–3948
12. Barakat MA (2010) Arab J Chem 4:361–377
13. Mayer-Gall T, Opwis K, Gutmann JS (2015) J Mater Chem A 3:386–394
14. Kumar PS, Sundaramurthy J, Sundarrajan S, Babu VJ, Singh G, Allakhverdiev SI, Ramakrishna S (2014) Energy Environ Sci 7:3192–3222
15. Pereao OK, Bode-Aluko C, Ndayambaje G, Fatoba O, Petrik LF (2016) J Polym Environ 2016:1–15
16. Peng S, Jin G, Li L, Li K, Srinivasan M, Ramakrishna S, Chen J (2016) Chem Soc Rev 45:1225–1241
17. Formhals A (1934) US patent 1,975,504
18. Doshi J, Srinivasan G, Reneker D (1995) Polym News 20:206–213
19. Fong H, Reneker DH (1999) J Polym Sci Part B Polym Phys 37:3488–3493
20. Reneker DH, Chun I (1996) Nanotechnology 7:216
21. Srinivasan G, Reneker DH (1995) Polym Int 36:195–201
22. Fong H, Chun I, Reneker DH (1999) Polymer 40:4585–4592
23. Lu X, Wang C, Wei Y (2009) Small 21:2349–2370
24. Lu X, Wang C, Favier F, Pinna N (2016) Adv Energy Mater 2016:1601301
25. Sill TJ, von Recum HA (2008) Biomaterials 29:1989–2006
26. Kim FS, Ren G, Jenekhe SA (2010) Chem Mater 23:682–732
27. Dai Y, Liu W, Formo E, Sun Y, Xia Y (2011) Polym Adv Technol 22:326–338
28. Wu J, Wang N, Zhao Y, Jiang L (2013) J Mater Chem A 1:7290–7305
29. Ray SS, Chen SS, Li CW, Nguyen NC, Nguyen HT (2016) RSC Adv 6:85495–85514
30. Ramaseshan R, Sundarrajan S, Jose R, Ramakrishna S (2007) J Appl Phys 102:7
31. Nalbandian MJ, Zhang M, Sanchez J, Choa YH, Nam J, Cwiertny DM, Myung NV (2016) Chemosphere 144:975–981
32. Ma Z, Ji H, Teng Y, Dong G, Zhou J, Tan D, Qiu J (2011) J Colloid Interf Sci 358:547–553
33. Jin R, Yang Y, Li Y, Yu X, Xing Y, Song S, Shi Z (2015) ChemPlusChem 80:544–548
34. Wang P, Du M, Zhu H, Bao S, Yang T, Zou M (2015) J Hazard Mater 286:33–544
35. Malwal D, Gopinath P (2016) RSC Adv 6:15021–115028
36. Li X, Zhao R, Sun B, Lu X, Zhang C, Wang Z, Wang C (2014) RSC Adv 4:42376–42382
37. Han C, Ma Q, Yang Y, Yang M, Yu W, Dong X, Wang J, Liu G (2015) J Mater Sci Mater Electron 26:8054–8064
38. Luo J, Luo X, Hu C, Crittenden JC, Qu J (2016) ACS Appl Mater Interfaces 8:18912–18921
39. Vu D, Li Z, Zhang H, Wang W, Wang Z, Xu X, Dong B, Wang C (2012) J Colloid Interf Sci 367:429–435
40. Wang H, Zhang P, Ma X, Jiang S, Huang Y, Zhai L, Jiang S (2014) J Hazard Mater 265:158–165
41. Li S, Yue X, Jing Y, Bai S, Dai Z (2011) Colloid Surf A 380:229–233

42. Mahapatra A, Mishra BG, Hota G (2013) J Hazard Mater 258:116–123
43. Hao L, Wang H, Cai R, Cheng J, Sun M, Li X, Jiang S (2016) RSC Adv 6:69947–69955
44. Deng S, Wang P, Zhang G, Dou Y (2016) J Hazard Mater 307:64–72
45. Xiong C, Li Y, Wang G, Fang L, Zhou S, Yao C, Chen Q, Zheng X, Qi D, Fu Y, Zhu Y (2015) Chem Eng J 259:257–265
46. Neghlani PK, Rafizadeh M, Taromi FA (2011) J Hazard Mater 186:182–189
47. Li X, Zhang C, Zhao R, Lu X, Xu X, Jia X, Wang C, Li L (2013) Chem Eng J 229:420–428
48. Zhao R, Li X, Sun B, Shen M, Tan X, Ding Y, Jiang Z, Wang C (2015) Chem Eng J 268:290–299
49. Wang J, Pan K, Giannelis EP, Cao B (2013) RSC Adv 3:8978–8987
50. Wang J, Pan K, He Q, Cao B (2013) J Hazard Mater 244:121–129
51. Zhao R, Li X, Sun B, Ji H, Wang C (2017) J Colloid Interf Sci 487:297–309
52. Dastbaz A, Keshtkar AR (2014) Appl Surf Sci 293:336–344
53. Kampalanonwat P, Supaphol P (2010) ACS Appl Mater Interfaces 2:3619–3627
54. Chaúque EF, Dlamini LN, Adelodun AA, Greyling CJ, Ngila JC (2016) Appl Surf Sci 369:19–28
55. Hong G, Li X, Shen L, Wang M, Wang C, Yu X, Wang X (2015) J Hazard Mater 295:161–169
56. Zhang C, Li X, Zheng T, Zhao R, Wang C (2014) Chem Res Chin Univ 30:685–689
57. Xiang T, Zhang Z, Liu H, Yin Z, Li L, Liu X (2013) Sci China Chem 56:567–575
58. Singh N, Balasubramanian K (2014) RSC Adv 4:27691–27701
59. Stephen M, Catherine N, Brenda M, Andrew K, Leslie P, Corrine G (2011) J Hazard Mater 192:922–927
60. Bahramzadeh A, Zahedi P, Abdouss M (2016) J Appl Polym Sci 133:42944
61. Zhijiang C, Jianru J, Qing Z, Haizheng Y (2015) RSC Adv 5:82310–82323
62. Zheng P, Shen S, Pu Z, Jia K, Liu X (2015) Fiber Polym 16:2215–2222
63. Hosseini M, Keshtkar AR, Moosavian MA (2016) Bull Mater Sci 39:1091–1100
64. Shariful MI, Sharif SB, Lee JJL, Habiba U, Ang BC, Amalina MA (2017) Carbohydr Polym 157:57–64
65. Aluigi A, Tonetti C, Vineis C, Tonin C, Mazzuchetti G (2011) Eur Polym J 47:1756–1764
66. Ma Y, Zhang B, Ma H, Yu M, Li L, Li J (2016) RSC Adv 6:30739–30746
67. Sharma DK, Li F, Wu YN (2014) Colloid Surf A 457:236–243
68. Wang X, Min M, Liu Z, Yang Y, Zhou Z, Zhu M, Chen Y, Hsiao BS (2011) J Membr Sci 379:191–199
69. Ki CS, Gang EH, Um IC, Park YH (2007) J Membr Sci 302:20–26
70. Wen HF, Yang C, Yu DG, Li XY, Zhang DF (2016) Chem Eng J 290:263–272
71. Zhao J, Li Z, Wang J, Li Q, Wang X (2015) J Mater Chem A 3:15124–15132
72. Wu C, Wang H, Wei Z, Li C, Luo Z (2015) Appl Surf Sci 346:207–215
73. Xin S, Zeng Z, Zhou X, Luo W, Shi X, Wang Q, Deng H, Du Y (2017) J Hazard Mater 324:365–372
74. Zeytuncu B, Akman S, Yucel O, Kahraman MV (2015) Water Air Soil Pollut 226:173
75. Sarioglu OF, Celebioglu A, Tekinay T, Uyar T (2016) Int J Environ Sci Tech 13:2057–2066
76. Li Y, Zhang J, Xu C, Zhou Y (2016) Sci China Chem 59:95–105
77. Huang CH, Hsieh TH, Chiu WY (2015) Carbohydr Polym 116:249–254
78. Liu H, Hsieh YL (2002) J Polym Sci Poly Phys 40:2119–2129
79. Celebioglu A, Uyar T (2011) Mater Lett 65:2291–2294
80. Tian Y, Wu M, Liu R, Li Y, Wang D, Tan J, Wu R, Huang Y (2011) Carbohydr Polym 83:743–748
81. Tong L, Chen J, Yuan Y, Cui Z, Ran M, Zhang Q, Bu J, Yang F, Su X, Xu H (2015) J Appl Polym Sci 132:41507
82. Rad LR, Momeni A, Ghazani BF, Irani M, Mahmoudi M, Noghreh B (2014) Chem Eng J 256:119–127
83. Habiba U, Afifi AM, Salleh A, Ang BC (2017) J Hazard Mater 322:182–194
84. Tan P, Wen J, Hu Y, Tan X (2016) RSC Adv 6:79641–79650

85. Najafabadi HH, Irani M, Rad LR, Haratameh AH, Haririan I (2015) RSC Adv 5:16532–16539
86. Irani M, Keshtkar AR, Moosavian MA (2012) Chem Eng J 200:192–201
87. Hota G, Kumar BR, Ng WJ, Ramakrishna S (2008) J Mater Sci 43:212–217
88. Aliabadi M, Irani M, Ismaeili J, Najafzadeh S (2014) J Taiwan Inst Chem E 45:518–526
89. Aliabadi M (2016) Fiber Polym 17:1162–1170
90. Razzaz A, Ghorban S, Hosayni L, Irani M, Aliabadi M (2016) J Taiwan Inst Chem E 58: 333–343
91. Hallaji H, Keshtkar AR, Moosavian MA (2015) J Taiwan Inst Chem E 46:109–118
92. Mahanta N, Valiyaveettil S (2013) RSC Adv 3:2776–2783
93. Taha AA, Wu YN, Wang H, Li F (2012) J Environ Manag 112:10–16
94. Cai J, Lei M, Zhang Q, He JR, Chen T, Liu S, Fu SH, Li TT, Liu G, Fei P (2017) Compos Part A 92:10–16
95. Xiao S, Ma H, Shen M, Wang S, Huang Q, Shi X (2011) Colloid Surf A 381:48–54
96. Sun B, Li X, Zhao R, Yin M, Wang Z, Jiang Z, Wang C (2016) J Taiwan Inst Chem E 62: 219–227
97. Xu GR, Wang JN, Li CJ (2012) Chem Eng J 198:310–317
98. Wang JN, Jia BB, Li CJ (2015) Nano 10:1550029
99. Guo H, Zhou H, Jin P, Li W, Ma Y, Koji I, Ji S (2016) RSC Adv 6:5965–5972
100. Luo C, Wang J, Jia P, Liu Y, An J, Cao B, Pan K (2015) Chem Eng J 262:775–784
101. Mirzabe GH, Keshtkar AR (2015) J Radioanal Nucl Chem 303:561–576
102. Taha AA, Qiao J, Li F, Zhang B (2012) J Environ Sci 24:610–616
103. Yari S, Abbasizadeh S, Mousavi SE, Moghaddam MS, Moghaddam AZ (2015) Process Saf Environ 94:159–171
104. Mokhtari M, Keshtkar AR (2016) Res Chem Intermed 42:4055–4076
105. Islam MS, Rahaman MS, Yeum JH (2015) Colloid Surf A 484:9–18
106. Wu S, Li F, Wang H, Fu L, Zhang B, Li G (2010) Polymer 51:6203–6211
107. Jia BB, Wang JN, Wu J, Li CJ (2014) Chem Eng J 254:98–105
108. Wu S, Li F, Xu R, Wei S, Wang H (2010) Mater Lett 64:1295–1298
109. Teng M, Wang H, Li F, Zhang B (2011) J Colloid Interf Sci 355:23–28
110. Sivakumar M, Mohan DR, Rangarajan R (2006) J Membr Sci 268:208–219
111. Bessbousse H, Rhlalou T, Verchère JF, Lebrun L (2008) J Membr Sci 307:249–259
112. Barakat MA, Schmidt E (2010) Desalination 256:90–93
113. Yalçın E, Gedikli S, Çabuk A, Karahaliloğlu Z, Demirbilek M, Bayram C, Sam M, Saglam N, Denkbaş EB (2015) Clean Techn Environ Policy 17:921–934
114. Zhou W, He J, Cui S, Gao W (2011) Fiber Polym 12:431–437
115. Aluigi A, Corbellini A, Rombaldoni F, Mazzuchetti G (2013) Text Res J 83:1574–1586
116. Lee D, Chen DWC, Chiu SF, Liu SJ (2015) Text Res J 85:346–355
117. Desai K, Kit K, Li J, Davidson PM, Zivanovic S, Meyer H (2009) Polymer 50:3661–3669
118. Liu Y, Ma H, Liu B, Hsiao BS, Chu B (2015) J Plast Film Sheet 31:379–400
119. Wu JJ, Lee HW, You JH, Kau YC, Liu SJ (2014) J Colloid Interf Sci 420:145–151
120. Zhao R, Li X, Sun B, Li Y, Li Y, Yang R, Wang C (2017) J Mater Chem A 5:1133–1144
121. Zhang M, Wang J, Li C (2015) Polym Eng Sci 55:421–428
122. Wang R, Guan S, Sato A, Wang X, Wang Z, Yang R, Hsiao BS, Chu B (2013) J Membr Sci 446:376–382

Chapter 6
Electrospun Filters for Organic Pollutants Removal

Anitha Senthamizhan, Brabu Balusamy, and Tamer Uyar

Abstract Increasing demand for access to clean and safe water around the globe emphasizes the development of new technologies for removing environmental pollutants. Especially, organic pollutants including dyes, volatile organic compounds (VOCs), polycyclic aromatic hydrocarbons (PAHs), pesticides, herbicides, and antibiotics prominently affect environmental health due to their hazardous nature. In the past several decades, advancements in electrospun fibrous membranes have resulted as an efficient filtering platform for removal of various pollutants in water, air, and soil. Electrospun nanofibers are efficient filters complementing their unique feature of accommodating a variety of functional molecules. The choice of material and the effect of experimental condition including pH, contact time, and adsorbent dosage on pollutant removal efficiency have been extensively reviewed previously. Our chapter focuses on recent progress in the developments of the electrospun functional nanofibrous composite membrane for various organic pollutants removal.

6.1 Introduction

The rapid global industrialization and modernization simultaneously have a negative impact on the society in terms of water pollution that continuously threatens the environmental health [1, 2]. Concerns were raised in the recent years for organic pollutants such as dyes, hydrocarbons, pesticides, fertilizers, phenols, plasticizers, biphenyls, detergents, oils, greases, and pharmaceuticals due to their carcinogenic nature and severe toxic impact on aquatic ecosystems [1, 3, 4]. The access to clean water is going to be a serious global challenge in the near future. Therefore, efficient removal of pollutants from water bodies is highly required. Up to date, several

A. Senthamizhan · T. Uyar (✉)
Institute of Materials Science and Nanotechnology, UNAM-National Nanotechnology
Research Center, Bilkent University, Ankara, Turkey
e-mail: senthamizhan@unam.bilkent.edu.tr; uyar@unam.bilkent.edu.tr

B. Balusamy
Italian Institute of Technology, Genova, Italy
e-mail: Brabu.Balusamy@iit.it

© Springer International Publishing AG, part of Springer Nature 2018   115
M. L. Focarete et al. (eds.), *Filtering Media by Electrospinning*,
https://doi.org/10.1007/978-3-319-78163-1_6

techniques (biologic, chemical, and physicochemical treatment) have been applied to remove the organic pollutants from wastewater in which the adsorption process is found to be an efficient method because of high adsorption capacity, simple design, ease operation, and scalability [5, 6].

So far, a variety of nanostructured materials were successfully established as adsorbents to remove the various pollutants in the environment [7–9]. In this regard, electrospinning is considered to be an efficient and convenient technique for producing nanofibrous membranes for filter application [10, 11]. These fibers generally possess extraordinary features including high surface area-to-volume ratios and tunable porosities, which make them excellent candidates for sensors, catalysts, and filtration fields [12–16]. Moreover, the fibrous structure promises the possible functionalization and exceptional accommodating features, resulting in superior composite materials that hold a great promise to remove the pollutants [17–19]. At the current scenario, it is sensible to describe the recent progress on the preparation of advanced functional fibrous materials for removal of organic pollutants. In this chapter, we will discuss the various routes to functionalize the electrospun fibrous membrane and their applications for removing potential organic pollutants such as dyes, volatile organic compounds (VOCs), polycyclic aromatic hydrocarbons (PAHs), etc., in the environment.

6.2 Electrospun Nanofibrous Filters for Organic Pollutants Removal

6.2.1 Removal of Dye Pollutants

Certainly, dyes are the best-known foremost pollutants of the environment due to their hazardous nature. So far, a variety of techniques were adopted for removal of dyes from water such as adsorption, coagulation, advanced oxidation, and membrane separation [20, 21]. Generally, the tiniest amount of dyes in water can be easily visualized and can be categorized in several ways as represented in Fig. 6.1, which depends on the structure, chromophores, and bonding nature [22]. The dye pollutant in water causes mutagenesis, carcinogenesis, chromosomal fractures, and respiratory toxicity even at lower concentrations.

Adsorption has been widely accepted as a promising technique due to its simplicity, efficiency, and environmental friendliness for removal of various pollutants in water [23–28]. The schematic depiction of removal of dyes using adsorbents is depicted in Fig. 6.2. The term adsorption refers to the accumulation of substances at the interface between two phases. So far, remarkable success has been achieved to design efficient adsorbents including carbon, silica gel, zeolites, clay, and alumina; also further efforts have been initiated to enhance their adaptability in advanced water treatment technologies. In general, it is difficult to recycle the nanoparticles

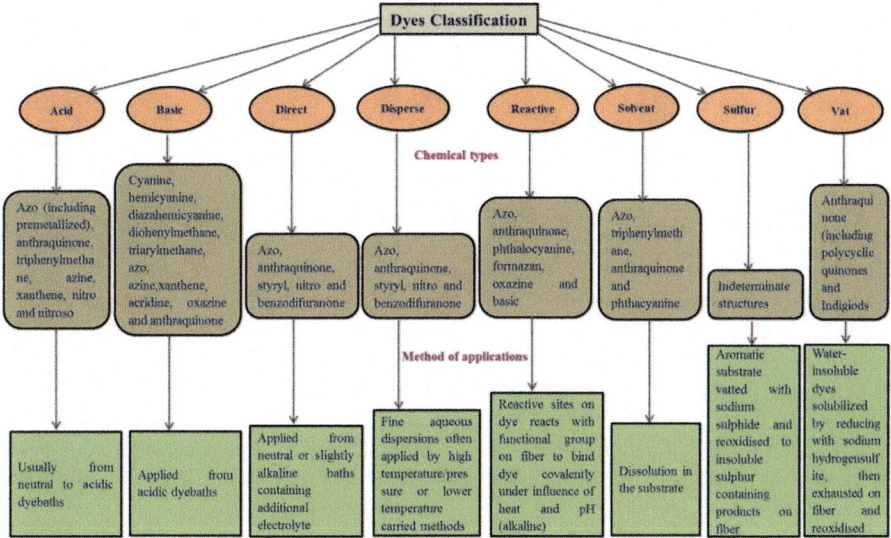

Fig. 6.1 Classification, applications, and method of application of dyes. Reproduced with permission from [22] © 2017 Elsevier

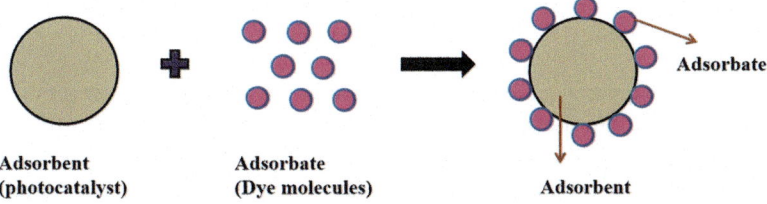

Fig. 6.2 Dye adsorption process using adsorbents. Reproduced with permission from [22] © 2017 Elsevier

after use which may lead to loss of activity and brings about the additional pollutant to the environment.

Therefore, development of new adsorbents combining the advantages of high adsorption efficiency, large surface area, and easy retrieval properties is considered to be of momentous importance in practical applications. As an example, γ-AlOOH (Boehmite) is an aluminum oxyhydroxide possessing plenty of OH groups on the surface resulting in an enhanced interaction with foreign molecules which makes it a suitable candidate for water decontamination. Miao et al. demonstrated the preparation of hierarchical SiO$_2$@γ-AlOOH (Boehmite) core/sheath fibers through the hydrothermal growth of Boehmite on the surface of electrospun SiO$_2$ fibers [29]. The schematic representation of the preparation procedure is given in Fig. 6.3.

At first, SiO$_2$ fibers were produced through the combination of sol–gel and electrospinning approach, followed by the removal of PVA (poly(vinyl alcohol)) via calcination. The high flexibility and thermal stability of electrospun SiO$_2$ fibers

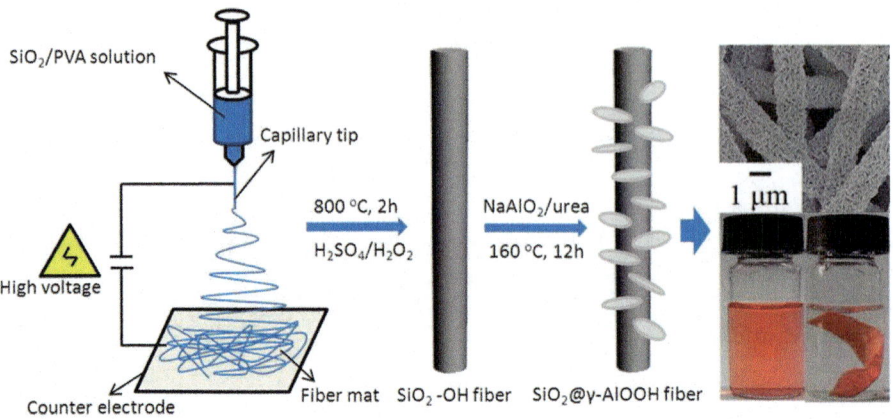

Fig. 6.3 Schematic illustration of the preparation of $SiO_2@\gamma$-AlOOH (Boehmite) core/sheath fibers and their dye removal performance. Reproduced with permission from [29] © 2012 American Chemical Society

facilitate the construction of hierarchical structure. The obtained open structure offers a much higher contact area between the adsorbent and pollutant in the water, which is considered to be a vital factor in determining the adsorption efficiency. The digital photograph of the $SiO_2@\gamma$-AlOOH (Boehmite) core/sheath fibrous membrane before and after adsorption of Congo red solution for over 90 min is depicted in Fig. 6.3.

The porous structured materials gained significant attention in adsorption of pollutants because of their enhanced surface area and pore volumes, resulting in high adsorption capacities [30, 31]. Usually, the adsorption process occurs in three steps: (1) adsorbate species are adsorbed onto the adsorbent surface through several forces such as hydrogen bonding, electrostatic interactions, van der Waals forces, hydrophobic interactions, etc.; (2) pollutants diffuse on the outer surface of the adsorbent and further enter into pores, if present; (3) finally, the pollutants are adsorbed onto the active site of the inner and outer surfaces of the adsorbent.

Figure 6.4a represents the adsorption isotherms of Methyl orange (MO) and Safranin O (SO) onto SiO_2 nanofibers. Figure 6.4b shows the schematic representation of dye adsorption onto SiO_2 nanofibers. The authors reported that mass of dyes adsorbed on the surface of nanofibers determines the adsorption kinetics. This might be due to sharing/exchange of electrons between dye and SiO_2 nanofibers molecules [32].

The results of Langmuir and Freundlich models suggested that the dyes are adsorbed on the SiO_2 nanofibers surface through the weak van der Waals forces. The calculated amounts of dyes were 714 and 958 (mg/g) for MO and SO, respectively. The negatively charged surface of the SiO_2 nanofibers favors the cation exchange of the dye which might be resulted in the high update of cationic dye. The lower adsorption of MO might be resulted from the presence of double bond character of the bridge in the azo group of MO that causes electron delocalization even for the acidic structures which establishes a potential barrier for adsorption. Im et al. dem-

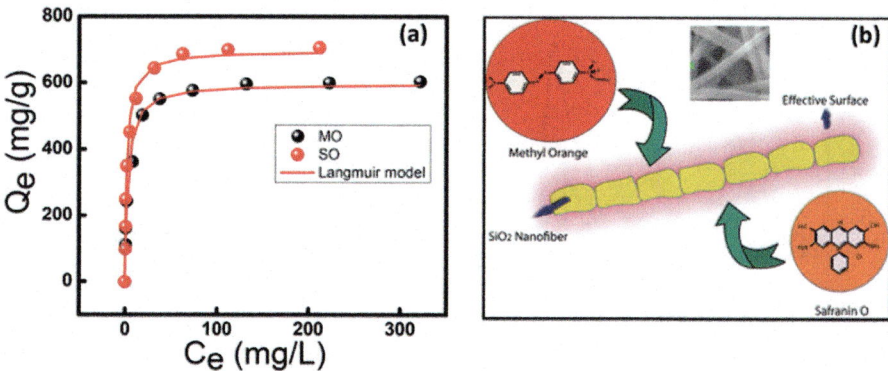

Fig. 6.4 (**a**) Adsorption isotherms for adsorption of MO and SO dyes onto SiO_2 nanofibers, (**b**) schematic representation of dyes adsorption onto SiO_2 nanofibers. Reproduced with permission from [32] © 2016 Elsevier

onstrated the graphene-embedded hydrogel nanofibers for detection and removal of aqueous-phase dyes [33].

To date, the outstanding characteristics of graphene have been extensively explored in diverse application. Even though, the agglomeration of graphene layers via π–π stacking and van der Waals interactions in the aqueous phase limits their enhanced performance. Considering all circumstances, authors introduced graphene into hydrogel fibers to take their benefits of high surface area, porosity, and hydrophilicity. The anchored, water-soluble polymer-based hydrogel architecture on graphene expedites an interaction between the graphene and aqueous-phase dyes. The prepared hydrogel nanofiber structure shows the appropriate mechanical properties enabling aqueous-phase applications, in which embedded graphene can interact more effectively with hydrophilic components. The adsorption capacities were as high as 0.43 and 0.33 mmol g^{-1} s^{-1} for methylene blue (MB) and crystal violet (CV), respectively, even in a 1.5 mL s^{-1} flow system. The adsorption capacities of graphene-embedded hydrogel nanofibers (GHNFs) for four dyes, namely, methylene blue (MB), crystal violet (CV), methyl orange (MO), and disperse red 1 (DR), were examined as a function of time.

In this regard, a piece of GHNF membrane was immersed in an aqueous dye solution and their corresponding ultraviolet (UV)–vis absorption spectra were taken at time intervals of 5 min (Fig. 6.5a–d). The results showed that GHNFs expressed diverse adsorption performances toward the dyes. The substantial decreases in the absorption peak intensities of MB and CV have been noted whereas no remarkable changes were observed for MO and DR. The changes in the dye concentration and adsorbed dye amount are plotted against time (Fig. 6.5e and f, respectively). The dye adsorption performance of GHNFs has been performed, and results showed that the removal efficiency of MB (93%) is slightly higher than that of CV (86%). Remarkably, GHNFs have no obvious effect toward MO (0%) and DR (4%). It is interesting to note that GHNFs exhibit selective adsorption of MB and CV.

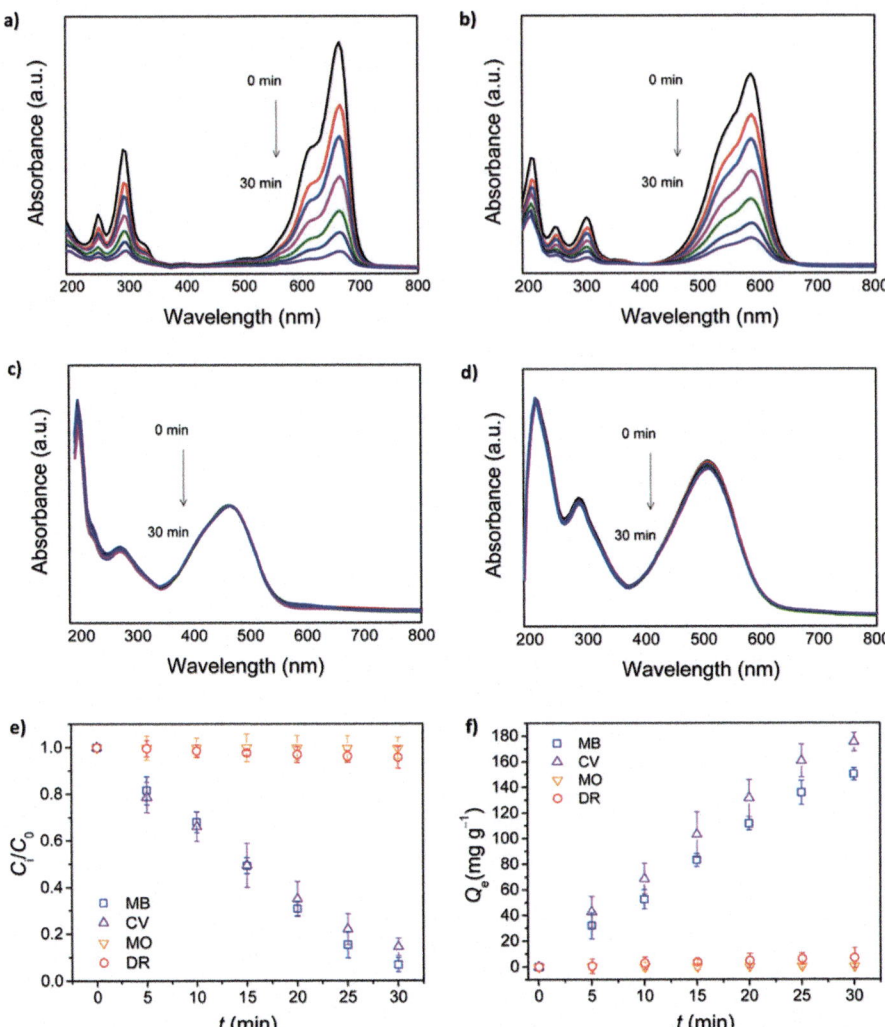

Fig. 6.5 Adsorption of dyes by graphene-embedded hydrogel nanofibers (GHNFs) in aqueous solution (at pH 6.0 and 25 °C in a closed cell). (**a–d**) Time-dependent changes (interval, 5 min) in UV–vis absorption spectra of dye solutions upon addition of GHNFs (0.1 g of GHNFs/50 mL of 1 mM dye solution): (**a**) methylene blue (MB), (**b**) crystal violet (CV), (**c**) methyl orange (MO), and (**d**) disperse red 1 (DR) (dissolved in 1/1 (v/v) ethylene glycol/water solution). Plots of the adsorption capacity versus time calculated from (**a–d**): (**e**) relative dye concentration change, in which C_0 and C_i denote the initial concentration and the instantaneous concentration at time t, respectively, and (**f**) the amount of dye adsorbed at time t, which was normalized by dividing it by the weight of GHNFs. Reproduced with permission from [33] © 2017 American Chemical Society

Alumina has been widely accepted as an excellent candidate for water purification. Even though the adsorption capacities are found to be relatively very high, their scalability is limited in their scattered forms like powder. Maintaining the porous structure of the adsorbent on the solid substrate brings a new way to enhance the adsorption performance. Wang et al. fabricated the mesoporous alumina-based adsorbent called silica/mesoporous alumina core–shell fibrous membranes (denoted as S/M fibrous membrane) in which silica is the core phase and mesoporous alumina is the shell phase [34]. Authors pointed that there are no reports for preparation of electrospun alumina fibers and their being used as adsorbents. Generally, the fragile nature of the alumina fibers cannot retain the membrane form even though they have a large surface area and high chemical and thermal stability. In this work, the authors paid more attention to the porous structure of the fibers and their integrity of the membrane form as well as scalability issues. Overall, the membrane demonstrates high mechanical strength (5.4 MPa) and high specific surface area, which might be attributed to the presence of dense silica core fibers and mesoporous alumina shell. The obtained silica/mesoporous alumina core–shell fibrous membranes exhibit good adsorption performance toward Congo red with an adsorption of 115 mg g^{-1} within 48 h. The integrality of the fibrous structure and their flexible nature have been well maintained after adsorption of dyes.

The reusable property of the core–shell membrane has been investigated, and results are shown in Fig. 6.6. The observed results showed that the initial adsorption capacity was 36.56 mg g^{-1}, which decreased slightly with increasing cycle times, and the adsorption capacity remained as high as 30.14 mg g^{-1}. The photograph shows the adsorption of Congo red for the fifth cycle, shown as an inset in Fig. 6.6. The stability of alumina shell has been determined by immersing the membrane in aqueous solution for several days, and then the elution of the Al content has been calculated as 0.0433 ppm. Overall, the results showed that the prepared membrane possesses high mechanical strength and good reusable performance in adsorption toward Congo red, as well as the high stability in aqueous solution of the membrane favors its application in the treatment of water.

Fig. 6.6 Relationship between the adsorption capacity and cycle times of the S/M core–shell fibrous membranes calcined at 700 °C. The inset is the optical image of the membranes after the fifth cycle of adsorption toward Congo red. Reproduced with permission from [34] © 2014 Royal Society of Chemistry

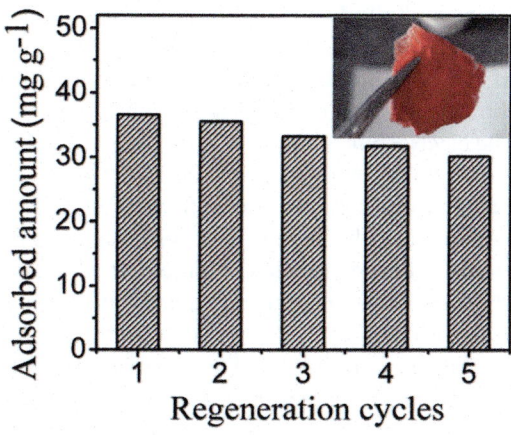

It has been already reported that the incorporation of nanoparticles into the polymer matrix significantly enhances the performance of the individual system as well as nanocomposites in diverse field of applications. Electrospinning approach has been accepted as a convenient technique for fabrication of polymer composites in which various nanostructured materials are successfully incorporated/decorated so far. In depth, the approach for making composites via electrospinning also affects the adsorption capabilities. Patel and Hota [35] investigated the adsorption capabilities of three differently prepared electrospun polyacrylonitrile (PAN)/iron oxide composite fibers using Congo red (CR) as model pollutant. In the first method, PAN/iron(III) acetylacetonate composite nanofibers prepared by electrospinning followed by a hydrothermal method for in situ growth of iron oxide nanoparticles on the surface of PAN nanofibers (named as PAN/IO(H) composite nanofibers). In the second method, the electrospun PAN nanofibers were immersed into the iron alkoxide solution followed by a hydrolysis reaction at 80 °C (named as PAN/IO(B)). In the third method, PAN/iron oxide composite nanofibers were prepared by blending the previously prepared iron oxide nanoparticles with the PAN solution followed by an electrospinning technique (PAN/IO(A). The schematic representation for the synthesis of PAN/iron oxide composite nanofibers membranes is shown in Fig. 6.7.

The proposed formation mechanism of iron oxide nanoparticles on the surface of the PAN nanofibers is presented in Fig. 6.8. The adsorption capacity has been studied for PAN/IO(H) and PAN/IO(A) composite nanofibers and can be explained as follows. It is reported that the point of zero charge (PZC) of iron oxide nanoparticles (12–100 nm) is in the range of 7.8–8.8. Therefore, the surface of the nanoparticles is positively charged at neutral pH. The result of which facilitates the electrostatic attraction between the positively charged iron oxide nanoparticles and the negatively charged CR dye molecule. Thus the higher adsorption capacities have been observed for PAN/IO(H) and PAN/IO(A) as compared to that of PAN/IO(B) and un-functionalized PAN nanofibers. Significantly, there is no obvious leaching of iron species into water and thereby this work suggests that the prepared composite fibers may have a possibility to serve as next-generation nanoadsorbents for the removal of CR dye owing to their simple preparation and high adsorption capacity.

As evidence, a molecular filter produced from incorporation of beta-cyclodextrin (β-CD) into polystyrene (PS) fibers through the electrospinning process has shown potential for removing organic compounds, that is, phenolphthalein (PhP). The electrospun nanofibers in this study were prepared from varying the concentration of both polymer PS (15–25% (w/v)) and β-CD (10–50% (w/w)). The filtration studies were conducted on PS and PS/CD fibers by immersing them into a PhP solution, and the change in absorbance of PhP was recorded as a function of time by UV–vis spectrometry. Figure 6.9a–d clearly illustrates that the absorbance of PhP solution decreased significantly over time in the presence of PS/CD fibers, due to the removal of PhP from solution by β-CD. It can be well understood from the experimental results that the distribution of CD at the surface of the fibers is seen in all PS–CD samples (Fig. 6.9e–g) and the distribution behavior will play a large role in determining the efficiency of the fibers as filters. The fact in the removal

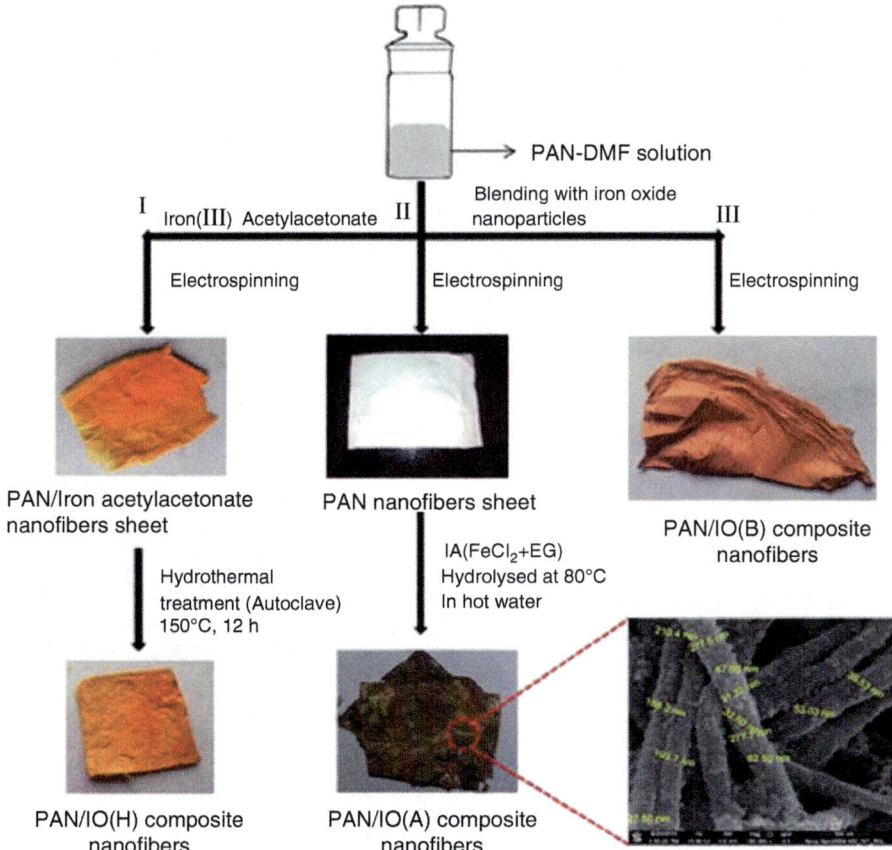

Fig. 6.7 Schematic representation for different synthetic methods of PAN/iron oxide composite nanofibers membranes. Reproduced with permission from [35] © 2016 Royal Society of Chemistry

efficiency solely depends on the distribution of CDs on the nanofibers by forming complexation with target molecules in the CD cavity, and the direct pyrolysis mass spectrometry studies confirmed the same. Hence, the study indicated the potential of using PS/CD fibrous membranes to filter organic molecules in purification/separation processes [36].

Later, Uyar et al. [37] conducted experiments to ensure the size effects and differences in affinity for selective inclusion complex (IC) formation with molecules; three different types of CDs (α-CD, β-CD, and γ-CD) were incorporated into electrospun PS nanofibers. Static time-of-flight secondary ion mass spectrometry (*static*-ToF-SIMS) analysis showed the presence of each type of CD on the PS nanofibers by the detection of both the CD sodium adduct molecular ions (M + Na+) and lower-molecular-weight oxygen-containing fragment ions. Further, comparative efficiency of the PS/CD nanofibers/nanoweb for removing phenolphthalein was determined by UV−vis spectrometry, and the kinetics of PhP capture was shown to

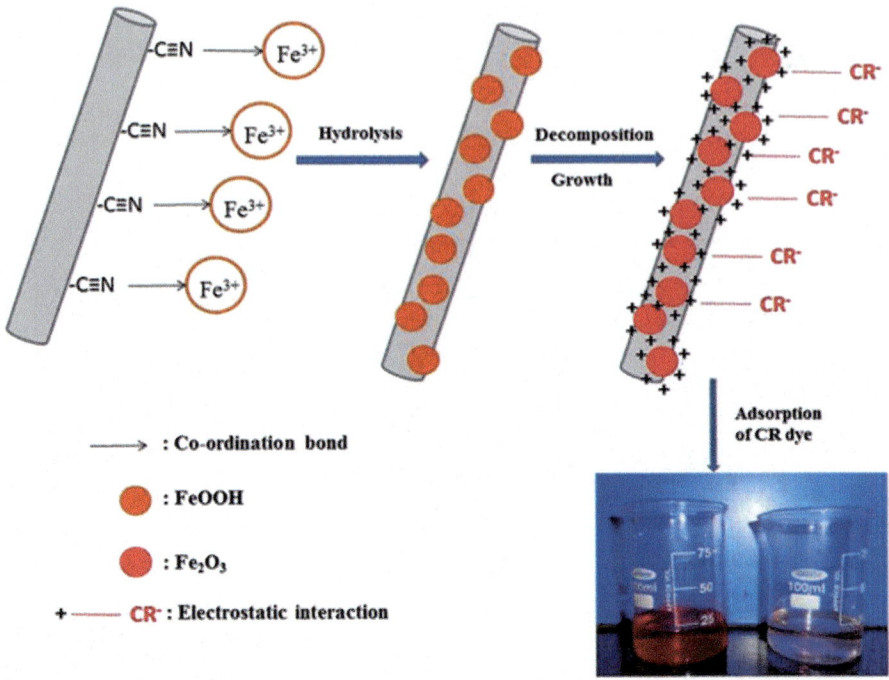

Fig. 6.8 Schematic representation of iron oxide nanoparticles grown on the surface of PAN nanofibers and mechanism of CR dye removal. Reproduced with permission from [35] © 2016 Royal Society of Chemistry

follow the trend PS/α-CD > PS/β-CD > PS/γ-CD. On the other hand, binding strengths of the PhP for the CD cavities showed the order of β-CD > γ-CD > α-CD. The study concluded that the cyclodextrins with different sized cavities can indeed filter organic molecules and can potentially be used for filtration, purification, and/or separation processes [37].

So far, a variety of materials have been proposed and successfully proven to be efficient adsorbents. Consequently, each of them has their own limitations. Considering the toxicity, recent research has focused on the development of cost-effective, renewable eco-friendly alternative bio-composites. Bioremediation is believed to be an alternative technology for decontamination of water systems by use of specific microorganisms such as bacteria, fungi, and algae. Usually, the usage of such microorganisms in bioremediation can be performed in two ways either as such or by immobilizing on a surface [38]. The immobilized microorganisms possess advantages over free cells in terms of their potential reusability, lower space and growth medium necessities, and higher resistance to environmental extremes. It is also well proven that the electrospun fibrous mat can act as an efficient platform for immobilization of specific microorganisms [39–42].

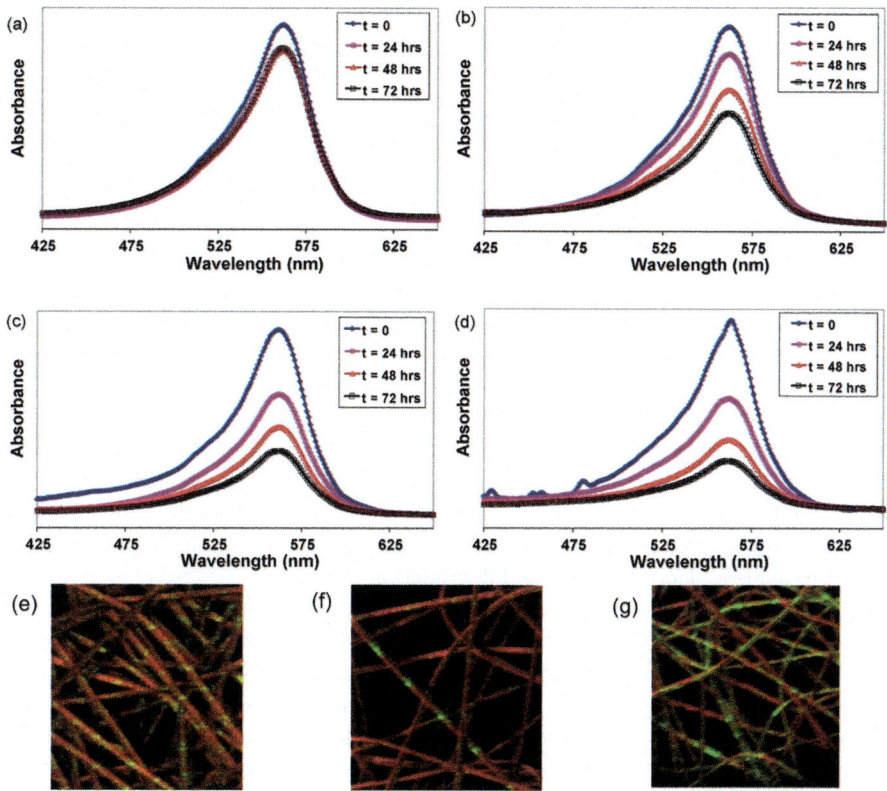

Fig. 6.9 UV–vis spectra of phenolphthalein (PhP) solution as a function of time after dipping the webs of (**a**) PS25, (**b**) PS20/CD10, (**c**) PS15/CD25, and (**d**) PS10/CD50. ToF-SIMS chemical images of fibers taken using the burst alignment mode. The distribution of CD on the surface of different PS/CD fibers shown by overlays of CD (green) on PS (red) fragment ions image, respectively (software color). (**e**) PS20/CD10, (**f**) PS15/CD25, and (**g**) PS10/CD50. Image area 75 μm × 75 μm. Reproduced/adapted with permission from [36] © 2009 Elsevier

San et al. demonstrated the three types of bacteria (*Aeromonas eucrenophila*, *Clavibacter michiganensis*, and *Pseudomonas aeruginosa*) immobilized cellulose acetate nanofibrous web (CA-NFW) for decolorization of methylene blue (MB) dye in aqueous medium [43]. The schematic representation of the preparation procedure for making nanofibers, attachment of bacteria cells, and their efficiency toward the decolorization of dye is shown in Fig. 6.10. The maximum dye adsorption capacity of bacteria-immobilized CA-NFWs is evaluated by studying the decolorization time (0–48 h) and different MB dye concentrations (20–500 mg L^{-1}). As a result, it is observed that the effective dye decolorization has been achieved within 24 h and 95% of the dye has been removed effectively. Thus, this work represents that the electrospun porous nanofibrous membranes are very effective solid supports for immobilizing bacterial cells.

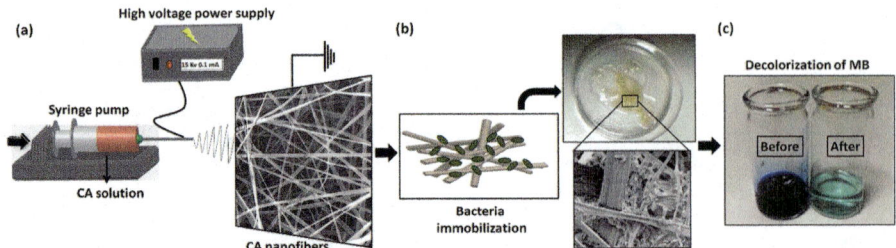

Fig. 6.10 Schematic representation of the electrospinning process for CA nanofibers, immobilization of bacterial cells on CA nanofibrous web, and photograph of the decolarization process. Reproduced with permission from [43] © 2014 Royal Society of Chemistry

Microalgae have been already identified as ideal biosorbents for wastewater treatment due to their specific nature such as ease of growth in simple medium, cheapness, availability, and high binding affinity. Yet, such a biosorbent shows limitations including low resistance to chemicals and heat. To overcome this problem, immobilization of microorganisms has been proposed and investigated as a viable means to achieve decolorization and degradation of dyes.

San Keskin et al. described the microalgae-immobilized polysulfone nanofibrous web (microalgae/PSUNFW) for removal of reactive dyes (Remazol Black 5 (RB5) and Reactive Blue 221 (RB221)) from wastewater [44]. They have selected *Chlamydomonas reinhardtii* as a model organism. The freestanding polysulfone nanofibrous web (PSU-NFW) was produced via electrospinning, and then *C. reinhardtii* was immobilized on their surface.

The SEM image of the electrospun PSU-NFW reveals that the obtained nanofibers are uniform and bead-free as represented in Fig. 6.11a–c. Figure 6.11d–f represents the photos of microalgae cells attached to the surface of PSU-NFW after 3 days and after 10 days from the start of the growth experiments. The green color of the membrane increased as concentration increases with respect to time. The attachment of the microalgae has been studied after 10 days of incubation. The observed results show that *C. reinhardtii* were strongly attached to the PSU-NFW. The decolorization rate for RB5 was calculated as 72.97 ± 0.3% for microalgae/PSU-NFW, whereas it was 12.36 ± 0.3% for the pristine PSU-NFW. In the case of RB221 solution, the achieved decolorization rates were 30.2 ± 0.23 and 5.51 ± 0.4% for microalgae/PSU-NFW and pristine PSU-NFW, respectively.

6.2.2 Removal of Volatile Organic Compounds (VOCs)

Environmental conservation has become a major social concern over few decades to avoid jeopardizing present natural resources. Volatile organic compounds (VOCs) are a large group of carbon-based molecules that easily evaporate at room temperature and considered as one of the major contributors to environmental pollution, emitted by a variety of sources including chemical and petrochemical industries,

Fig. 6.11 Representative SEM images of (**a**) pristine PSU-NFW and (**b** and **c**) microalgae/PSU-NFW (under different magnifications). Photos (**d–f**) of attachment of free microalgae cells on PSU-NFW for 3- and 10-day attachment processes. Reproduced/adapted with permission from [44] © 2015 American Chemical Society

vehicles, ships, aircrafts, etc. [45–49], which undoubtedly has severe adverse effects on human and environmental health. It has been well reported that VOCs are capable of causing headache, nausea, mutagenic and carcinogenic effects [50–52], also has more adverse effect on lungs followed by pulmonary exposure [53]. Therefore, an enormous effort has been put forward to control the pollution arising from VOCs in environment. In this context, various methods have been developed including adsorption, oxidation, biological treatment, distillation, etc., for removal of VOCs from air and water [49, 54–59]. Son very recently discussed different methods adopted for controlling VOCs as illustrated in Fig. 6.12 [49]. However, many complications exist in applying them in actual field due to many influencing factors, thus attempts on identifying new methods and materials continue without hiatus.

As discussed earlier, nanotechnology deals with control of matters at a nanoscale level and production of materials with specific properties and functionalities that have shown potential applications in various sectors. Among the various available techniques for preparation of fibers, electrospinning has been recognized as a simple, versatile, and cost-effective method capable of producing nanofibers from a variety of polymers, inorganic and organic-inorganic compounds. The nanofibers produced through electrospinning have received great attention over the past decade for their application in healthcare to environment due to their specific properties like high surface-to-volume ratio, porous nature, permeability, and fiber diameter [60–66]. The electrospun nanofibers also significantly contributed to controlling VOCs

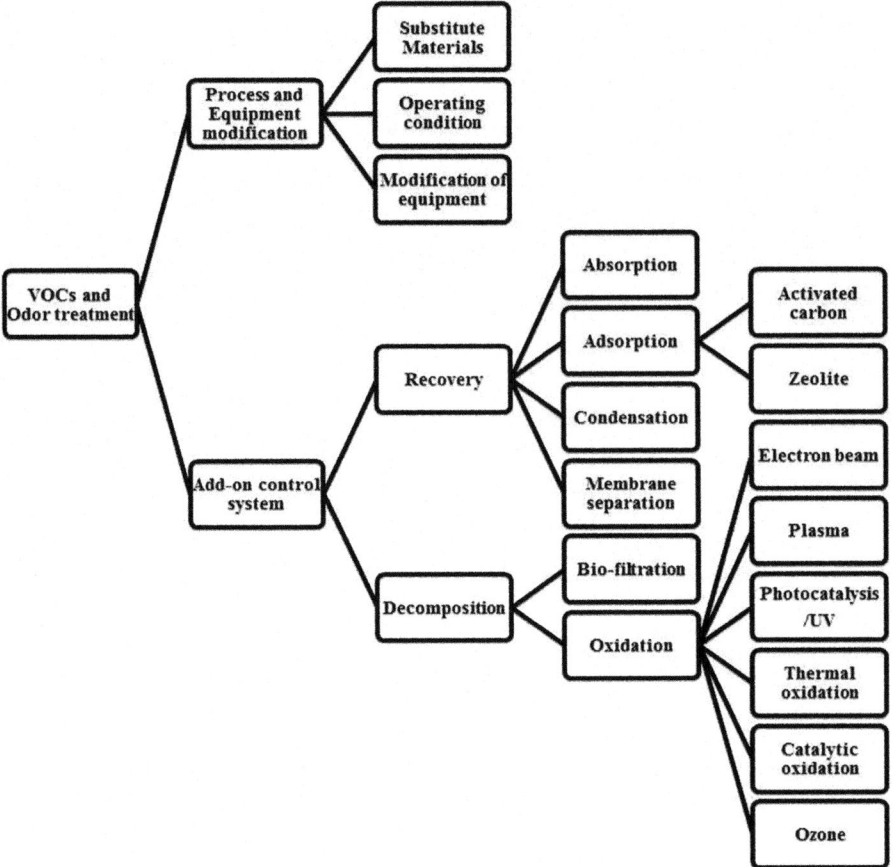

Fig. 6.12 Techniques for treatment of VOCs and odorous compounds. Reproduced with permission from [49] © 2017 Elsevier

pollution in environment using various approaches. This part of this chapter discusses about the different nanofibers intended in removal of VOCs.

A recent study conducted by Chu et al. [67] compared the adsorption/desorption performance of VOCs on electrospun nanofibers prepared from polystyrene (PS) nanofibers, acrylic resin (AR) nanofibers, and polystyrene–acrylic resin (PS–AR) composite nanofibers with activated carbons (ACs), carbon nanotubes (CNTs), and activated carbon fibers (ACFs) against n-butanol, toluene, chlorobenzene, anisole, nitrobenzene, dichloromethane, ethyl acetate, benzene, acetone, and n-hexane [67]. The surface characteristics of the fibers were investigated by using SEM and TEM; corresponding images are presented in Fig. 6.13. The diameters of the fibers were measured as 300–800 nm for the PS nanofibers, 200–400 nm for the AR nanofibers, and 200–900 nm for the PS–AR nanofibers. A homemade experimental setup has been adapted to conduct the adsorptive experiment using 5 mg of the

Fig. 6.13 SEM and TEM images of (A, A′) PS nanofibers, (B, B′) AR nanofibers, and (C, C′) PS-AR nanofibers. Reproduced with permission from [67] © 2015 Royal Society of Chemistry

sample followed by gas chromatography measurement. Further, desorption and regenerative performances of the samples were investigated under thermal conditions. The outcome of the study indicated that nanofibers possess selective adsorption property against VOCs based on their composition. The adsorption efficiency of the nanofibers is comparable to that of ACFs and CNTs determined per weight unit. Interestingly, the electrospun nanofibers showed better thermal desorption properties than ACs, CNTs, and ACFs, which emphasize the advantage of the nanofibers over other adsorbents due to gas adsorption/desorption which occurs at the surface of the nanofibers. Overall results of this study found that the electrospun nanofiber could be a potential VOC adsorbent since it holds certain selectivity, short adsorption equilibration time, preferable adsorption/desorption efficiency, reusability, and temperature effect.

The electrospun polyacrylonitrile (PAN) carbon nanofibers were reported for their applicability as novel alternative adsorbents with commercial ACFs, A–10 for volatile organic compounds (VOCs) removal. The results of isosteric enthalpy of adsorption and adsorption energy distribution tests/equations prove the higher adsorption capacities of PAN carbon nanofibers than the commercial ACFs for benzene removal [68]. In another study, activated carbon nanofibers (ACNFs) produced from electrospinning of polyacrylonitrile solutions demonstrated higher adsorption efficiency of VOCs (benzene and ethanol) than activated carbon fibers (ACFs) in a static vapor adsorption system. The ACNFs were prepared by initially electrospinning 10 wt% PAN solution (ACNF10-800) and 15 wt% solution (ACNF15-800) followed by stabilizing the electrospun nanofibers by heating the samples to 280 °C at a rate of 1 °C min⁻¹ in air and kept at 280 °C for 2 h, then the fibers were further heated up and carbonized at 800 or 850 °C for 30 min in nitrogen with a flow rate of 212 mL min⁻¹ controlled by the rotameter. Further, the samples were activated by adding 30 vol.% steam-controlled nitrogen gas at the same temperature for 10, 30, and 50 min. The burn-offs were calculated based on the weights of the stabilized electrospun fibers and the resultant samples and denoted as ACNF15-850-w30, 70, and 80. On the contrary, ACFs were prepared from oxidized PAN fibers by carbonization at 850 °C for 30 min and then activated at the same temperature with nitrogen flow containing 50 vol.% steam for 30 min. Outcome of the study indicated that ACNFs fabricated by the method of electrospinning demonstrated higher adsorption capacities for VOCs than ACFs at extremely low relative pressures. Adsorption isotherms of benzene and ethanol were measured at 20 °C using a volumetric adsorption system. The ACNFs demonstrated higher adsorption capacities for VOCs than ACFs at extremely low relative pressures. The physical and chemical properties of ACNFs have significantly improved by varying the activation conditions. The increased burn-off conditions facilitate the enhanced microporosity in the nanofibers which results in better adsorption performance as shown in Fig. 6.14. In addition, surface chemistry has an important effect on the adsorption of polar VOCs. The ACNF with higher oxygen content possesses a higher adsorption capacity for ethanol [69].

Electrospun polyurethane fibers fabricated through the electrospinning method by using polyurethanes based on 4,4-methylenebis(phenylisocyanate) (MDI) and aliphatic isophorone diisocyanate as the hard segments and butanediol and tetramethylene glycol as the soft segments were electrospun from their solutions in *N,N*-dimethylformamide. The electrospun polyurethane fibers had an average diameter of 2 μm and were found to be smooth and nonporous in nature as illustrated in Fig. 6.15. The fibers were then studied for their adsorbent efficiency against VOCs such as toluene, hexane, and chloroform. The adsorption efficiencies of activated carbon were also studied for comparison. The sorption capacity of the polyurethane fibers was found to be similar to that of activated carbon specifically designed for vapor adsorption, even though the activated carbon possessed a many-fold higher surface area. Additionally, the polyurethane fibers demonstrated a complete reversible adsorption and desorption using a simple purging with nitrogen at room temperature, which was not possible in the case of activated carbon. The fibers possessed a high affinity toward toluene and chloroform, but aliphatic hexane lacked the necessary strong attractive interactions with the polyurethane chains and therefore was less strongly absorbed [70].

Another study reported by Feng et al. [71] revealed the removal of chloroform by using the electrospun poly(vinylidene fluoride) (PVDF) nanofiber membrane in a membrane air-stripping system. Gas-stripping membrane distillation system used in this study is shown schematically in Fig. 6.16. In brief, the results of the study displayed that owing to high surface hydrophobicity and appropriate pore sizes of the electrospun nanofiber the VOC was effectively removed. The overall mass transfer coefficient of chloroform through the nanofiber membrane was found to be

Fig. 6.14 Adsorption/
desorption isotherms of (**a**)
benzene and (**b**) ethanol on
three samples, and (**c**)
adsorption isotherms of
benzene and ethanol on
ACNF15-850-w80 at
20 °C. Reproduced with
permission from [69] ©
2013 Springer

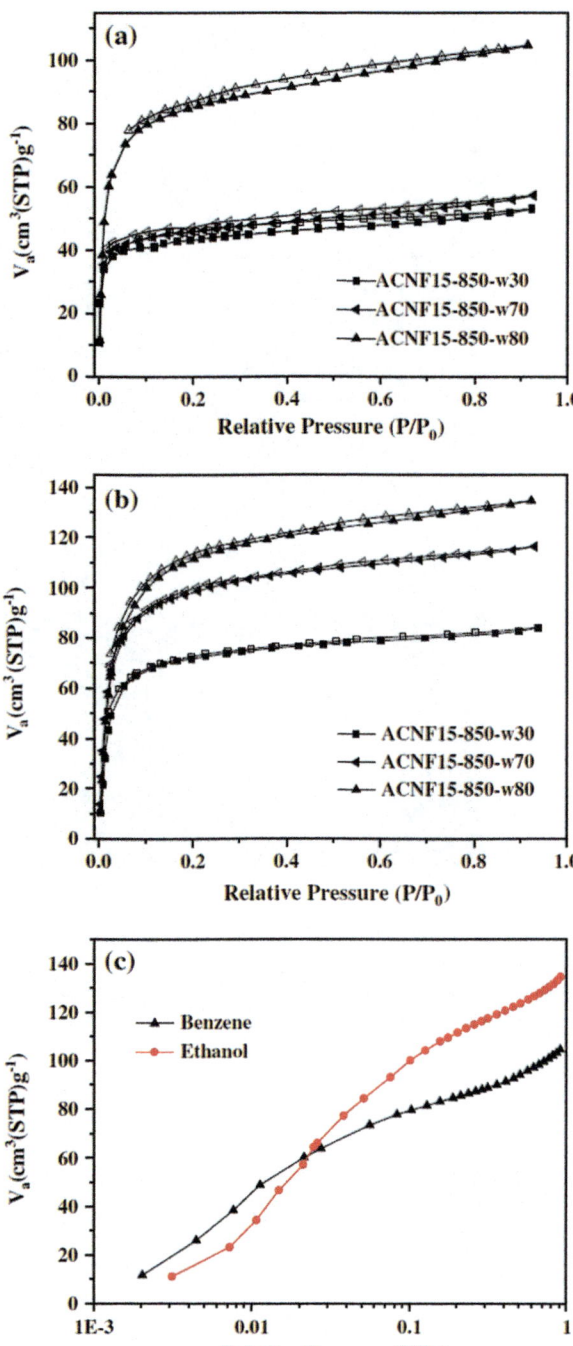

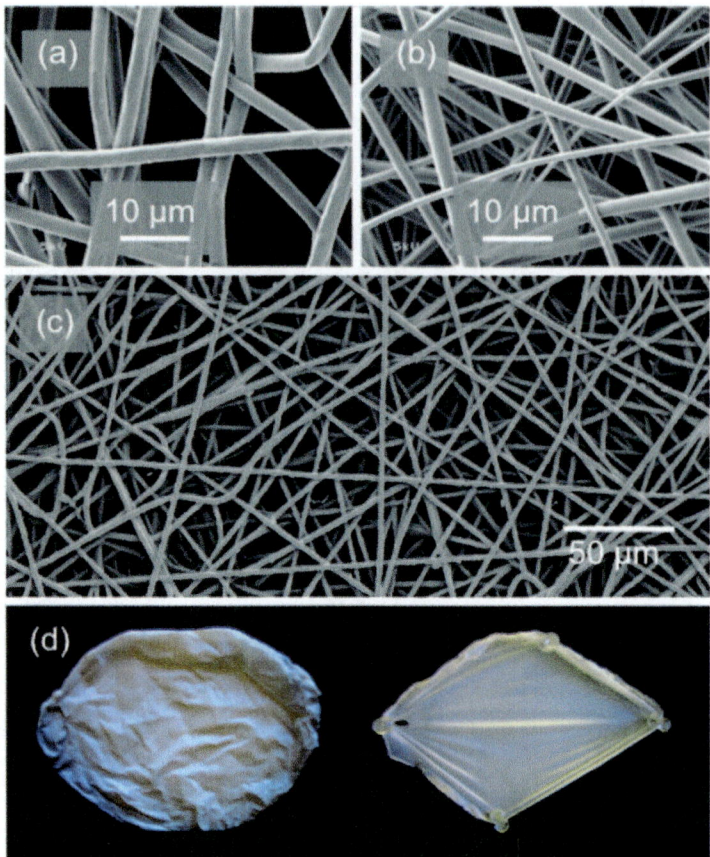

Fig. 6.15 SEM images of (**a**) MDI-based and (**b**) isophorone-based nonwoven fiber mats showing smooth, nonporous fiber surfaces. (**c**) SEM image showing uniformity of MDI-based PU fiber diameter and mat density. (**d**) Strength and elasticity of mats demonstrated by lack of tearing or breaking of stretched fiber mat held in place with push pins. Reproduced with permission from [70] © 2011 American Chemical Society

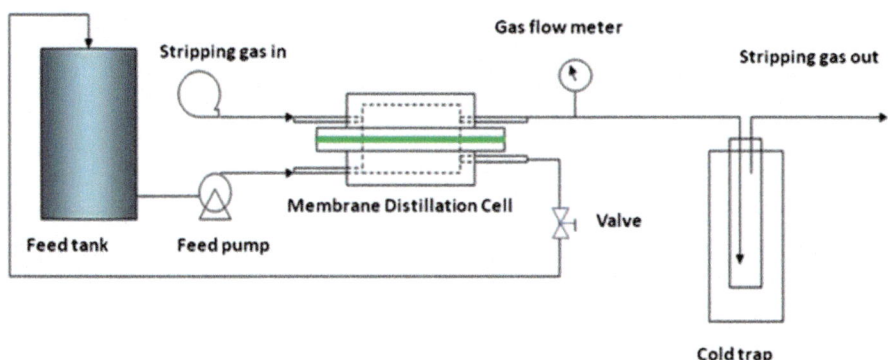

Fig. 6.16 Gas-stripping membrane distillation system. Reproduced with permission from [71] © 2011 Elsevier

2.40×10^{-5} m/s, and this value was higher than the values obtained for a membrane air-stripping system based on a hollow fiber module. Notably, activation energy of the overall mass transfer coefficient was found to be nearly equal to that of the Henry constant for chloroform absorption in water. Hence, the study concludes that the use of the electrospun nanofiber contributed to effectively remove the VOC because of their unique properties [71].

Nevertheless, functionalization of nanofibers with various compounds could improve the efficiencies of nanofibers. Hence, several attempts were made to decorate or incorporate different compounds for enhancing the peculiar performances. In the consideration of current discussion, to enhance the surface oxygen content Guo et al. [72] added graphene oxide (GO) sheets at weight percentages (3 and 6 wt.%) into polyacrylonitrile (PAN), further the composite fibers were prepared through the electrospinning process. Pristine carbon nanofibers without addition of GO (ACNF) and conventional PAN-based activated carbon fiber (ACF) were also used for the sake of comparison. The resultant nanofibers were tested for their performances in adsorption of benzene and butanone through vapor adsorption equilibrium isotherms. Interestingly, addition of GO increased mesopores and high surface oxygen content of the nanofibers and exhibited higher adsorption capacities toward both the VOCs as compared to the pristine carbon nanofiber and commercial ACF, as shown in Fig. 6.17. Additionally, surface oxygen content greatly influenced the adsorption behavior of composite nanofibers as butanone, a polar compound, had higher adsorption rate than the nonpolar benzene. The highest benzene and butanone adsorption capacities of the prepared composite nanofibers at 20 °C and with relative pressure of 0.98 reached 83.2 and 130.5 cm^3 g^{-1}, respectively. In addition, adsorption capacity ratio isotherms of butanone relative to benzene on the three nanofibers at 20 °C were also calculated against various relative pressures and displayed butanone possessed higher adsorption than benzene. Obviously, VOC adsorption performance of the series of composite nanofibers was significantly improved by the introduction of GO with increasing concentration. In brief, a change of approach in preparing electrospun nanofibers evidently demonstrated the effective adsorption performances against conventional fibers [72].

In another approach aimed for enhancing the removal performance of nanofibers toward pollutants, cyclodextrins (CDs) incorporation has been explored as an ideal strategy. The CDs are cyclic oligosaccharides comprising of 1,4-linked glucopyranoside units with either six, seven, or eight glucose units organized in a cyclic structure, named as α-, β-, and γ-cyclodextrins, respectively. Our group has extensively investigated the influence of filtering and removal capacity of VOCs upon functionalization/incorporation of cyclodextrin (CD) molecules in electrospun nanofibers. As evidence, Uyar et al. [73] reported the preparation of beta-cyclodextrin (β-CD) incorporated poly(methyl methacrylate) (PMMA) nanofibers using electrospinning in order to develop functional nanofibrous webs for organic vapor waste treatment. The systematic characterization studies confirmed distribution of β-CD molecules on the nanofiber surface. In order to investigate the entrapment of aniline, styrene, and toluene by the PMMA nanowebs containing 10%,

Fig. 6.17 Adsorption equilibrium isotherms of (**a**) benzene and (**b**) butanone on various samples at 20 °C. Normalized (**c**) benzene and (**d**) butanone adsorption isotherms of various samples with respect to specific surface area at 20 °C. Reproduced with permission from [72] © 2016 Elsevier

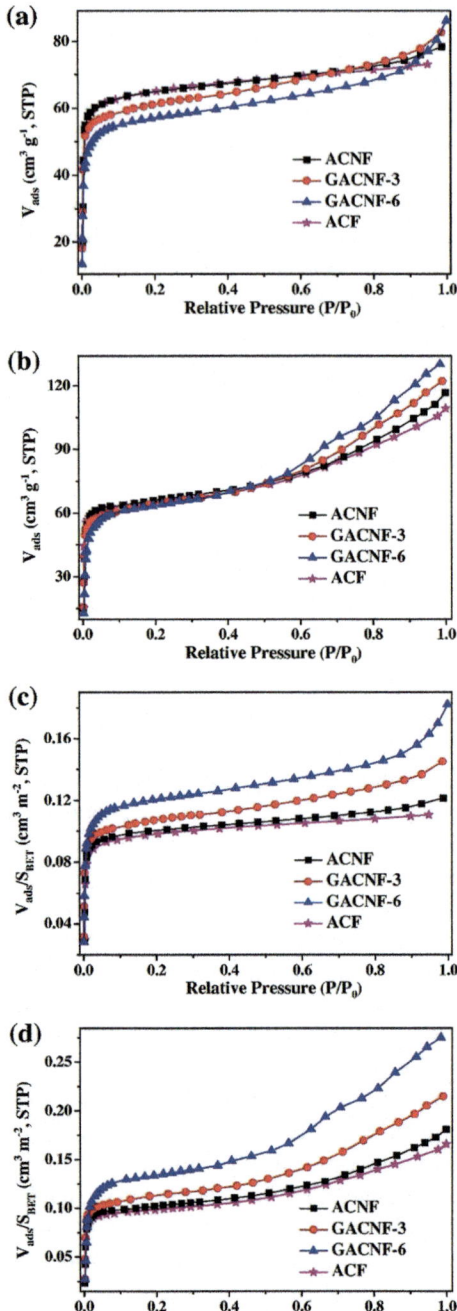

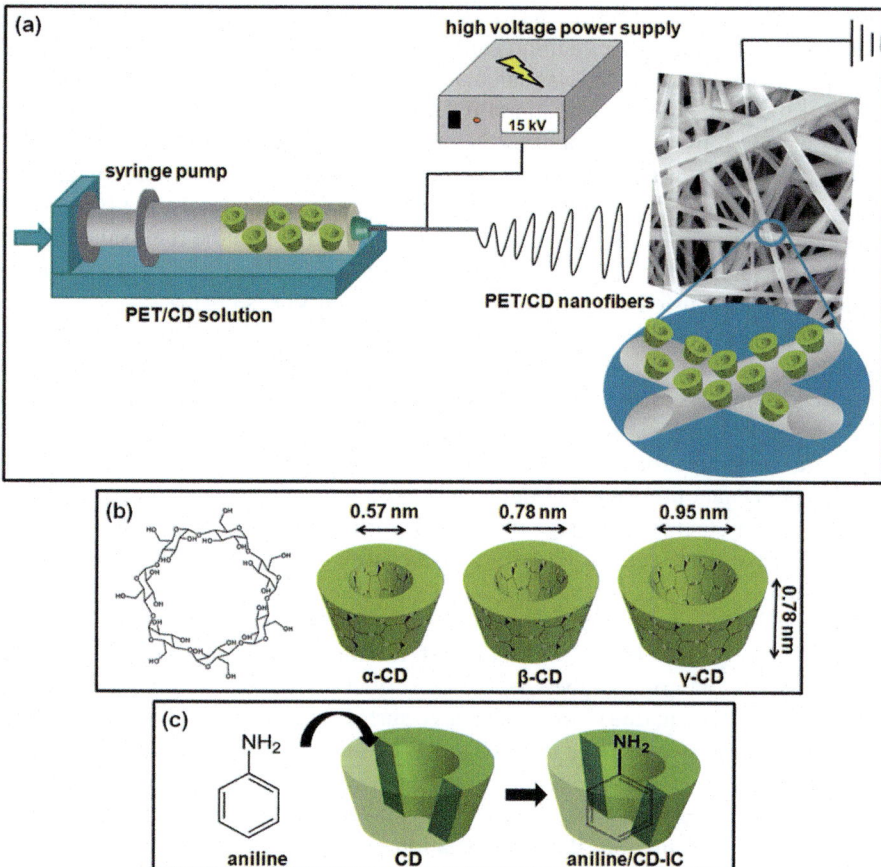

Fig. 6.18 Schematic representations of (**a**) electrospinning of PET/CD solution, (**b**) chemical structure of β-CD and approximate dimensions of α-CD, β-CD, γ-CD, and (**c**) formation of aniline/CD-IC. Reproduced with permission from [74] © 2014 John Wiley and Sons

25%, and 50% β-CD (PMMA/β-CD10, PMMA/β-CD25, and PMMA/β-CD50), the evolution profiles of single-ion pyrograms of the characteristic PMMA- and β-CD-based products were investigated. The pyrolysis mass spectrometry results clearly indicated that the β-CD functionalized nanofibers effectively trap organic waste vapors with respect to increase in exposure period and the amount of CD present in the nanowebs [73].

Kayaci and Uyar [74] incorporated α-CDs, β-CDs, and γ-CDs into polyester (PET) nanofibers by the electrospinning process. Interestingly, bead-free PET/CD nanofibers were obtained from lower polymer concentration owing to the fact that CDs incorporation increased spinnability of electrospun solution. A schematic representation shown in Fig. 6.18 clearly demonstrates the electrospinning process and formation of inclusion complexes with aniline. The entrapment performance of the

PET and PET/CD nanofibrous webs was investigated by using aniline vapor as a model VOC. The amount of aniline entrapped by PET, PET/α-CD, PET/β-CD, and PET/γ-CD nanofibrous webs was calculated by HPLC. The absorption was approximately 1300 ppm aniline by pristine PET nanofibers, whereas PET/α-CD and PET/β-CD nanofibrous webs captured approximately 2600 ppm and PET/γ-CD nanofibrous webs captured approximately 3400 ppm at the end of entrapment test. The findings of the study suggest that CD functionalized electrospun PET nanofibers can be promising filtering materials for air filtration and the removal of VOCs due to very high surface area of nanofibrous webs and surface-associated CD molecules having inclusion complexation capability with VOCs [74].

Another study reported the preparation of functional nylon 6,6 nanofibers incorporating cyclodextrins (CD) with enhanced thermal stability. The nanofibers were prepared by incorporating three types of α-CD, β-CD, and γ-CD at different ratios (25% and 50%, w/w, with respect to nylon 6,6). The resultant membranes without CD were ineffective for entrapment of toluene vapor from the environment, whereas nylon 6,6/CD nanofibrous membranes can effectively entrap toluene vapor from the surrounding by taking advantage of the high surface-to-volume ratio of nanofibers with the added advantage of inclusion complexation capability of CD presenting on the nanofiber surface. Particularly, the nanofibers incorporating 50% of β-CD showed enhanced entrapment performance than α-CD and γ-CD in both concentrations as can be seen from Fig. 6.19a. In brief, it is understood that CD functionalized electrospun nylon 6,6 nanofibers would be very effective for the removal of VOCs from the environment due to their very large surface area along with inclusion complexation capability of surface associated CD on the nanofibers. The modeling studies for formation of an inclusion complex between CD and toluene were also performed by using ab initio techniques and the results confirmed that based on the first principles calculation, the complexation energy of toluene-CD is higher for β-CD compared to α-CD and γ-CD as illustrated in Fig. 6.19b [75].

Celebioglu and Uyar [76] electrospun gamma cyclodextrin (γ-CD) nanofibers from a DMSO–water solvent system without using any carrier polymeric matrix and further investigated the molecular entrapment capability toward aniline and toluene. The γ-CD nanofibers were placed in a desiccator saturated with aniline or toluene vapor and the measurements of organic vapors were performed by ^{1}H-NMR. The γ-CD powder was treated under the same environment for comparison purpose. The outcome of molecular entrapment experiment indicated that γ-CD nanofibers were quite successful in entrapping of VOCs (aniline and toluene) by inclusion complexation, whereas γ-CD in powder form did not show any entrapment capability. The overall findings suggest that these electrospun CD nanofibers can be used as molecular filters and/or nanofilters for the removal of organic volatile compounds (VOCs) from the environment [76].

Very recently, the molecular entrapment performance of polymer-free nanofibers of hydroxypropyl-beta-cyclodextrin (HPβCD) and hydroxypropyl-gamma-cyclodextrin (HPγCD) was investigated toward two common VOCs (aniline and benzene). The encapsulation efficiency of CD samples was investigated by varying different factors including CD form (NF and powder), electrospinning solvent (DMF and water), CD

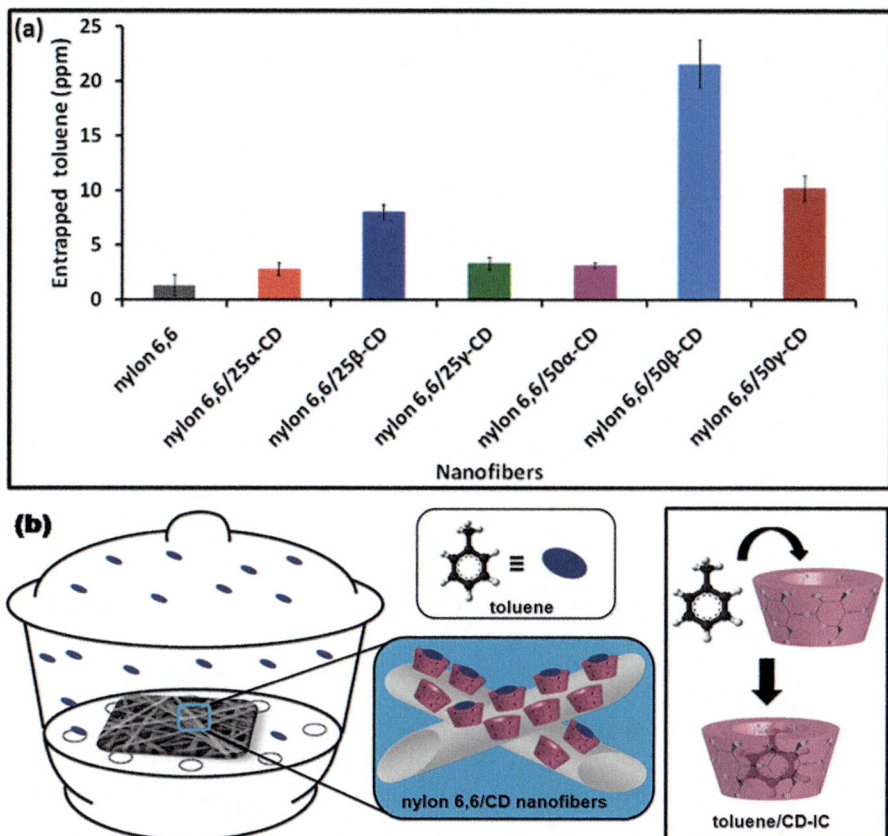

Fig. 6.19 (**a**) The amount of entrapped toluene (ppm) by the electrospun nanofibers and (**b**) schematic representation of the formation of toluene/CD-IC. Reproduced/adapted with permission from [75] © 2015 John Wiley and Sons

(HPβCD and HPγCD), and VOC (aniline and benzene) types. The observed results recommended that CD NF could entrap a higher amount of VOCs from surroundings than their corresponding powders. Besides, molecular entrapment efficiency of CD NF depended on CD, solvent, and VOC types. Further, the inclusion complexation ability of CD molecules was based on very high surface area and versatile features of CD NF. So evidently, all these reports proved that incorporation of CDs into electrospun nanofibers serves as useful filtering material for the purposes VOCs removal [77].

Likewise, many other attempts have been made to incorporate different compounds into electrospun nanofibers for their enhanced performances in VOCs removal. For instance, commercially available fly ash particles (FAPs) at different concentrations (0, 10, 30, 50, and 70 wt%) were blended with polyurethane (PU) and electrospun to obtain fly ash/polyurethane composite fibrous membrane. The fibers were tested against different VOCs, including chloroform, benzene, toluene, o-xylene, and styrene. The experimental data showed that, among all VOCs, styrene

Fig. 6.20 Cyclic absorption behavior of different VOCs on a PU composite mat obtained from 30 wt% FAPs containing PU solution. Reproduced with permission from [78] © 2013 Elsevier

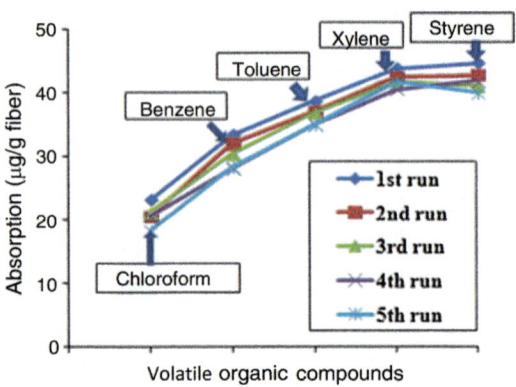

was the highly absorbed one, regardless of composition of PU fibers. PU fibers with 30 wt% FAPs showed enhanced VOC absorption capacity, which is 2.52–2.79 times (for five VOCs) higher compared to that of pristine PU fibers. To ensure the practical applicability of the composite fibers in order to their reusability performance, PU fibers containing 30 wt% FAPs have been studied for repeated VOC adsorption experiment. The recyclability results demonstrated that composite fibrous mats have nearly equal absorption efficiency up to five cycles as depicted in Fig. 6.20 [78].

Following the same trend, Loess powder (LP), which is generally used to adsorb the pollutants in water as nanoparticles (NPs), has been incorporated into PU nanofibers at 0, 10, 30, and 50 wt% concentrations. The outcome of adsorption experiments revealed that the PU/LP nanofibers containing 30 wt% LP NPs has highest VOC absorption capacity and follows the trend toluene > benzene > chloroform [79]. In another recent study reported by Ge and Choi [80], polyurethane/rare earth (PU/RE) composite nanofibrous membranes were produced by electrospinning and were intended for the application of removing VOCs from air. The VOC adsorption studies of styrene, xylene, toluene, benzene, and chloroform had shown PU nanofibers containing 50% RE powder highly adsorbed styrene [80].

6.2.3 Removal of Polycyclic Aromatic Hydrocarbons (PAHs)

Polycyclic aromatic hydrocarbons (PAHs) are other vital environmental pollutants comprising two or more fused benzene rings that are arranged in different configurations and released into environment from natural and anthropogenic sources. The chemical structures of few commonly studied PAHs are given in Fig. 6.21 [81]. PAHs consisting of fewer than four rings are classified as low-molecular-weight compounds, whereas high-molecular-weight compounds consisting of four or more rings are usually colored, crystalline solids at ambient temperature, poorly soluble in water, and highly mobile in the environment because of their physicochemical

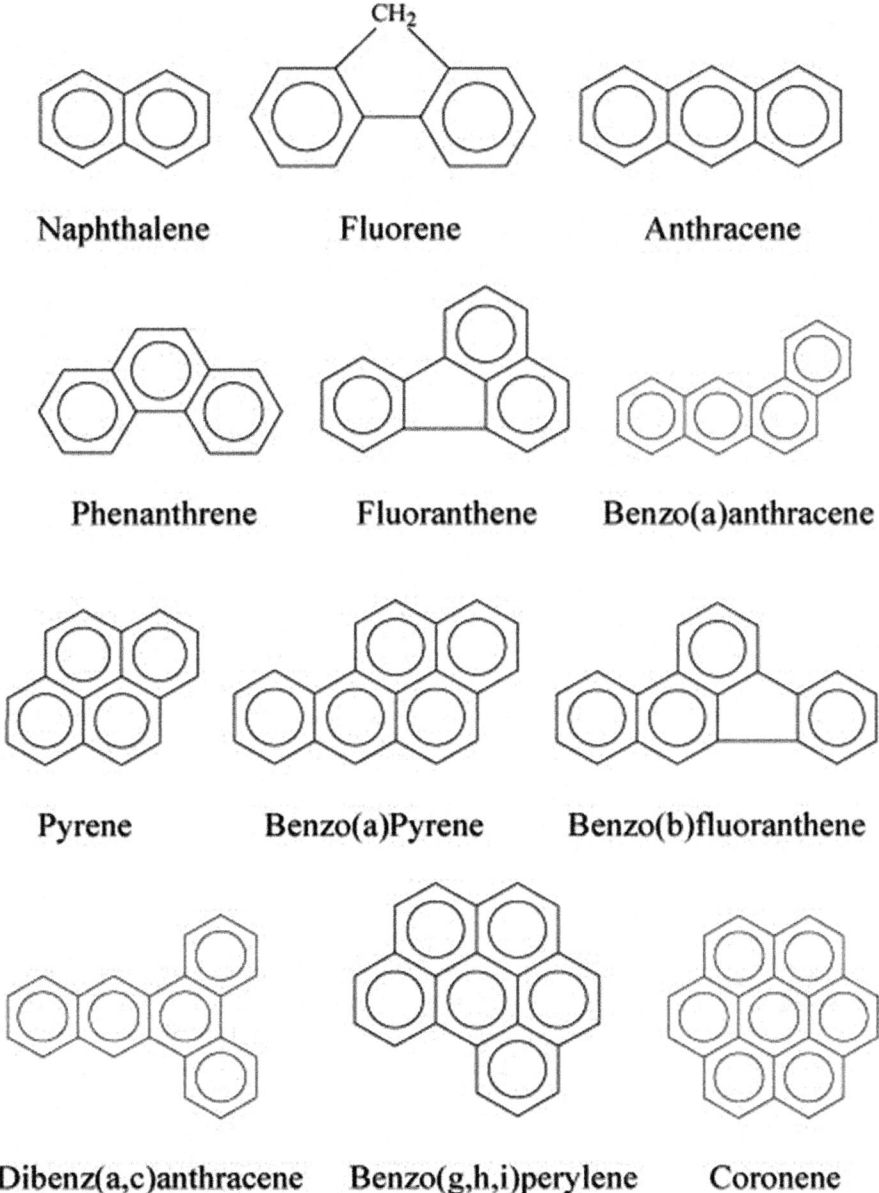

Fig. 6.21 Chemical structures of some commonly studied PAHs. Reproduced with permission from [81] © 2009 Elsevier

properties and distributed across air, water, and soil [81–83]. Due to their ubiquitous persistence in environment, negative effects of PAHs are well documented for a long time. PAHs cause increased risk of skin, bladder, lung, and gastrointestinal cancers. Moreover, reactive molecules of some PAHs tend to bind with cellular proteins and DNA molecules resulting in cell damage, mutation, malformation development, and cancers. The deleterious effects of PAHs on environment and human health have been reviewed in detail and reported [84–86]. Hence, a variety of methods, including bioremediation, have been exploited to control the PAHs pollution in environment till now and continue to discover for better solution [81, 87, 88]. A variety of nanomaterials, including carbon nanotubes, graphene, and iron nanoparticles, have been reported for their efficiency in significant removal of PAHs [89–91]. In addition, electrospun nanofibers proved for their potency in removing PAHs that will be discussed in the following section.

Electrospun nanofibers of poly(styrene-*co*-methacrylic acid), poly(styrene-*co*-p-styrene sulfonate), and polystyrene have been used as solid-phase extraction (SPE) sorbents to extract six aromatic hydrocarbon compounds, including nitrobenzene, 2-naphthol, benzene, *n*-butyl *p*-hydroxybenzoate, naphthalene, *p*-dichlorobenzene in water samples. The high surface-to-volume ratio of PS nanofibers ensured a strong interaction between the nanofiber sorbent and aromatic hydrocarbons and resulted in higher sensitivity, reproducibility, detection limits, and reducing the time. The applicability of the sorbents was also investigated for real-time water samples [92]. A study conducted by Dai et al. [93] used five types of nanofibrous membranes prepared by electrospinning poly(ε-caprolactone) (PCL), poly(D,L-lactide) (PDLLA), poly(lactide-*co*-caprolactone) (P(LA/CL)), poly(D,L-lactide-*co*-glycolide) (PDLGA), and methoxy polyethylene glycol-poly(lactide-*co*-glycolide) (MPEG-PLGA) for removal of anthracene (ANT), benz[a]anthracene (BaA), and benzo[a]pyrene (BaP) from aqueous solution. All the sorption processes were mostly achieved within the first 30 min of reaction, and then became more gradual until the equilibrium was reached after 3 h. The sorption rates of the PCL ENFMs for PAHs were relatively faster than others, and their sorption capacity for ANT, BaA, and BaP reached 75.6, 78.3, and 73.5 $\mu g\ g^{-1}$ at 60 min, respectively. The differences in sorption of these three pollutants toward ENFMs are based on the hydrophobic interactions, hydrogen bonding interactions, π–π bonding interaction, and pore-filling factors. A detailed representation of the sorption mechanism is presented in Fig. 6.22 [93].

In order to enhance the adsorption and degradation of PAHs, three different types of electrospun nanofibers poly(D,L-lactide) (PDLLA), poly(D,L-lactide-*co*-glycolide) (PDLGA), and methoxypolyethylene glycol–poly(lactide-*co*-glycolide) (MPEG–PLGA) immobilized with laccase (LCEFMs) have been prepared and used for the removal of phenanthrene, fluoranthene, benz[a]anthracene, and benzo[a]pyrene from the soil sample collected at Baisha Shoal (China) in the aqueous solution. The PAH removal studies indicated the degradation efficiencies for the four PAHs were ranked in the order of phenanthrene > fluoranthene > benz[a]anthracene > benzo[a] pyrene. Specifically, the degradation efficiencies of the PAHs by all three LCEFMs

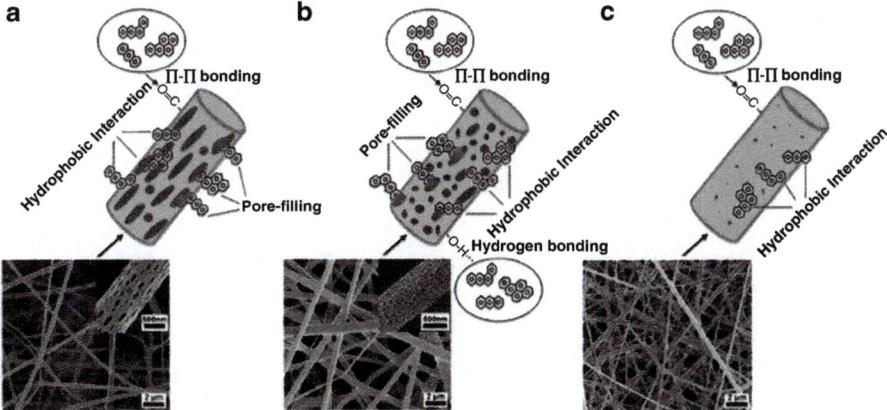

Fig. 6.22 Schematic diagram of the sorption of three PAHs on PDLLA ENFMs (**a**); MPEG-PLGA ENFMs (**b**); P(LA/CL), PCL, and PDLGA ENFMs (**c**). Reproduced with permission from [93] © 2011 Elsevier

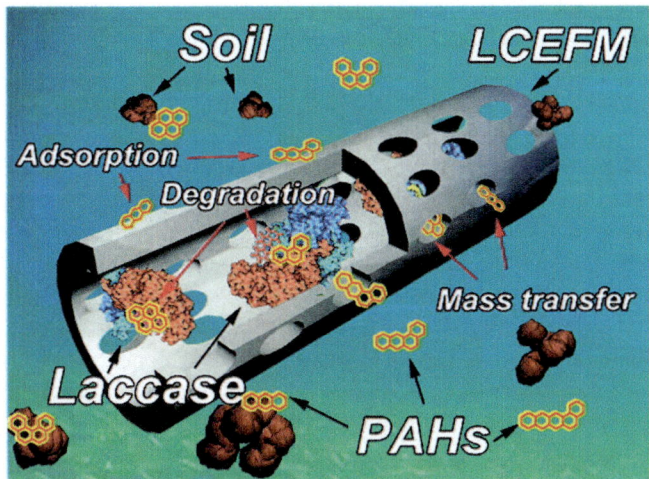

Fig. 6.23 The scheme of the PAH phase transfer and degradation process by laccase-carrying electrospun fibrous membrane. Reproduced with permission from [94] © 2011 American Chemical Society

were much higher than those by free laccase, especially for benzo[a]pyrene, whose degradation efficiency by free laccase was less than 30%, whereas those by the three LCEFMs exceeded 70%. Moreover, the PAHs in aqueous solution was adsorbed on the surface of the LCEFMs and concentrated around the active sites of laccase; thus, the degradation rates of PAHs were significantly enhanced and the corresponding process is explained in Fig. 6.23 [94]. Later, the approach has been adopted to remove the PAHs in contaminated water [95].

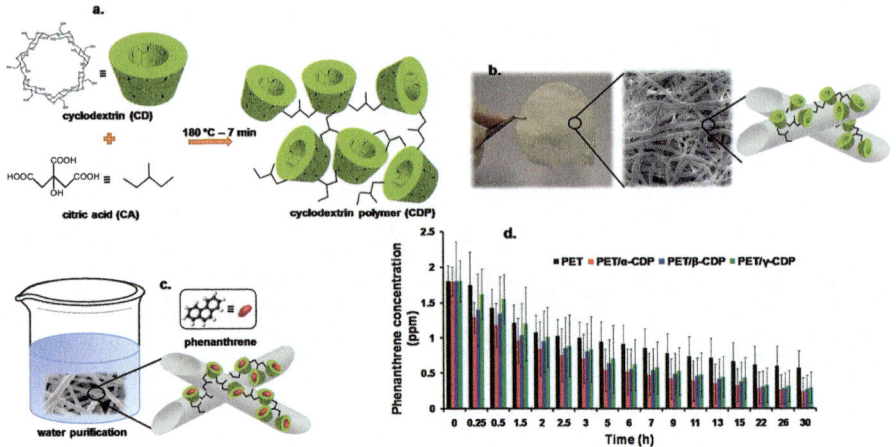

Fig. 6.24 Schematic representations of (**a**) formation mechanism of CDP and (**b**) the representative photograph of PET/CDP nanofibrous mat and its SEM image and schematic representation of PET/CDP nanofibers. (**c, d**) Time-dependent decrease of phenanthrene concentration in the aqueous solution containing nanofibrous mats and phenanthrene complex formation. Reproduced/adapted with permission from [96] © 2013 Elsevier

Considering the beneficial nature of cyclodextrins in capturing environmental pollutants, electrospun polyester (PET) nanofibers with cyclodextrin polymer (CDP) were produced (PET/CDP) with all three α-CD, β-CD, and γ-CD types and investigated for their efficiency in removing the PAH compound phenanthrene from aqueous solution and, further, their efficiency was compared to that of pristine PET nanofibers. The schematic representation of nanofiber preparation, mechanism, and their characteristics is shown in Fig. 6.24. In this study, PET nanofibers were obtained by electrospinning of 22.5% (w/v) PET solution in TFA/DCM (50/50, v/v). The chemical reaction cannot occur between cyclodextrin (CD)/citric acid (CTR) and PET nanofibers directly, since PET, a polymer based on terephthalic acid and ethylene glycol, does not contain free reactive groups.

Therefore, we modified the surface of the electrospun PET nanofibers through the polymerization reaction between CTR and CD. The mechanism of the CDP formation is schematically described in Fig. 6.24a and b. The molecular filtration studies revealed that the concentration of phenanthrene in the aqueous solution decreased within the contact time as depicted in Fig. 6.24c and d. The adsorption of phenanthrene by PET nanofibers for the first 2 h was observed, and then the concentration of phenanthrene slightly decreased over time. On the other hand, the decrease of phenanthrene concentration for PET/CDP mats was more significant. Although less amount of PET/CDP nanofibers was used compared to that of PET nanofibers for filtration test, the removal efficiency of phenanthrene from its aqueous solution was better due to structure of CDP structure that plays a crucial role in molecular capturing of phenanthrene. Therefore, the functionalization approach on electrospun nanofibers proved to change the PET/CDP as a molecular filter for water purification through complexation of the phenanthrene with CDP [96].

Cellulose acetate (CA) nanofibers grafted with beta-cyclodextrin (β-CD) prepared by combining electrospinning and "click" reaction have been demonstrated for their efficiency in removing the PAH compound phenanthrene from aqueous solution. In brief, at first, β-CD and electrospun CA nanofibers were modified as azide-β-CD and propargyl-terminated and then "click" reaction was performed between modified CD molecules and CA nanofibers to attain permanent grafting of CDs on nanofibers surface (see Fig. 6.25). The morphological behavior of CA nanofibers before and after the surface modification has been compared by SEM as depicted in Fig. 6.25b. As it is shown in the SEM images, some changes occurred at the morphology of CA nanofibers after each process. The functionalized nanofibers (CA-CD) were capable of removing high amounts of PAH than the CA nanofibers because of the repulsive interactions between the hydrophobic guest and the aqueous environment, and more favorable interactions between hydrophobic guest and apolar CD cavity as depicted in Fig. 6.25c [97].

6.2.4 Removal of Other Organic Pollutants (Antibiotics and Pesticides)

Besides, pollution owing to antibiotics in environment has raised another major concern because of great risk even at trace levels. To date, a variety of technologies (ion-exchange, membrane filtration, adsorption, and photocatalytic degradation) have been used to remove antibiotics from aqueous solution [98–100]. There are few studies that reported the removal of antibiotics by using electrospun nanofibers, as an example, novel Fe_3O_4/polyacrylonitrile (PAN) composite nanofibers (NFs) were prepared by a two-step process, namely, electrospinning and solvothermal methods. The composite fibers were evaluated to check their feasibility in antibiotic removal using tetracycline (TC) as the model antibiotic molecule by batch adsorption experiments. The results showed that Fe_3O_4/PAN composite NFs were effective in removing TC with no impactful loss of Fe in the pH regime of environmental interest (5–8) and their adsorption capacity calculated from the Langmuir isotherm model was 257.07 mg g^{-1} at pH 6. From the experimental analysis, it was proposed that the adsorption of TC on the composite membranes occurs mainly through the cation exchange and complex formation between the Fe atom from the composite NFs and the deprotonated moieties from the TC molecule, as depicted in Fig. 6.26 [101].

Organophosphates (OPs) are a group of organic phosphorous compounds and mostly used as pesticides/insecticides and are of much concern because of their potency to bind with acetylcholinesterase through dermal exposure, resulting in disturbance of nervous impulses and inhibiting nerve cell functions [102]. Various methods to degrade OPs similar to other pollutants were explored, including electrochemical, biodegradation, photolytic, catalytic oxidation, and destructive adsorp-

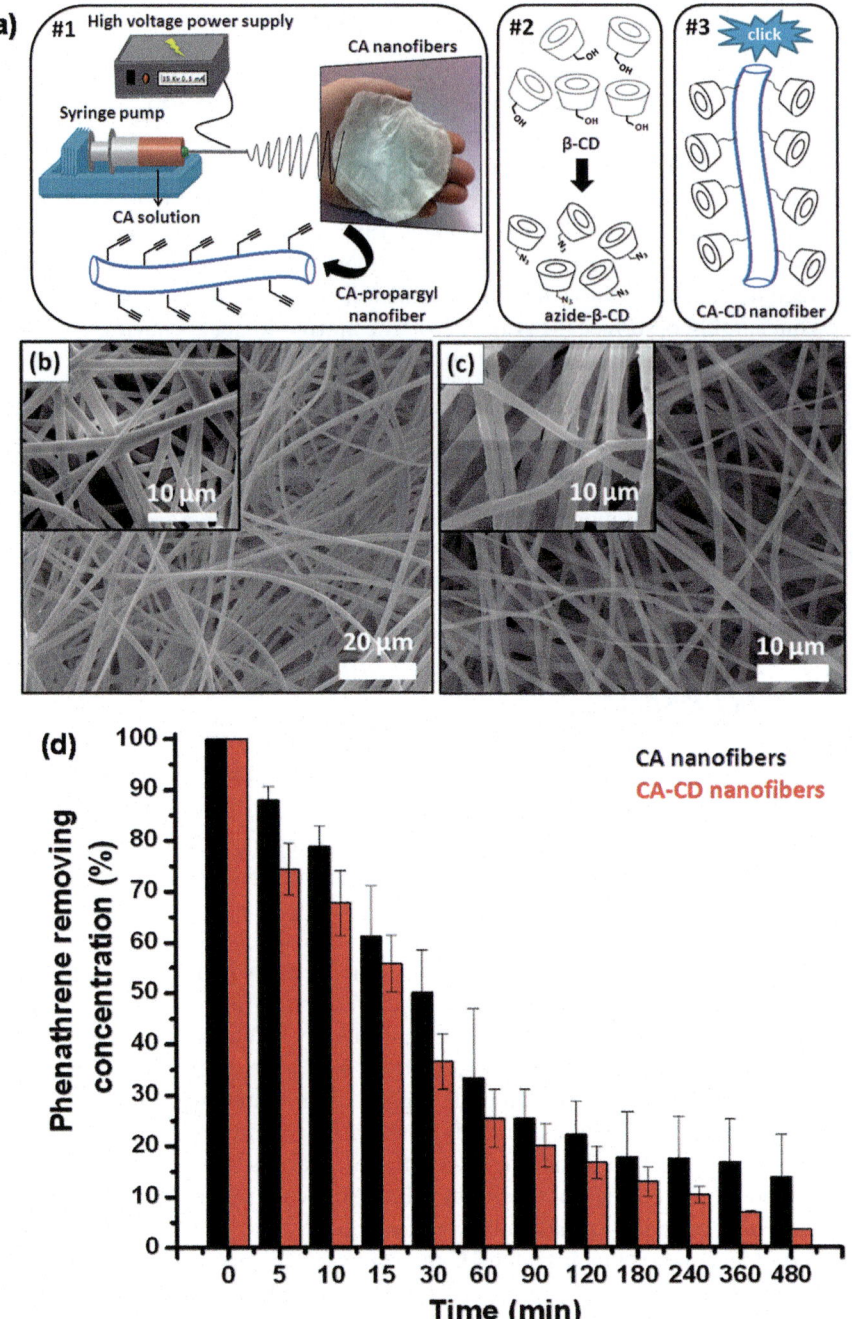

Fig. 6.25 (**a**) Schematic representation of electrospinning of CA nanofibers and modification of CA-propargyl nanofibers with azide-β-CD by "click" reaction. (**b, c**) Representative SEM images **Fig. 6.25** (continued) of CA-propargyl and CA-CD nanofibers, respectively. The insets show higher magnification images. (**d**) The time-dependent decrease of phenanthrene concentration in aqueous solution which contains CA and CA-CD nanofibers webs. Reproduced/adapted with permission from [97] © 2014 Elsevier

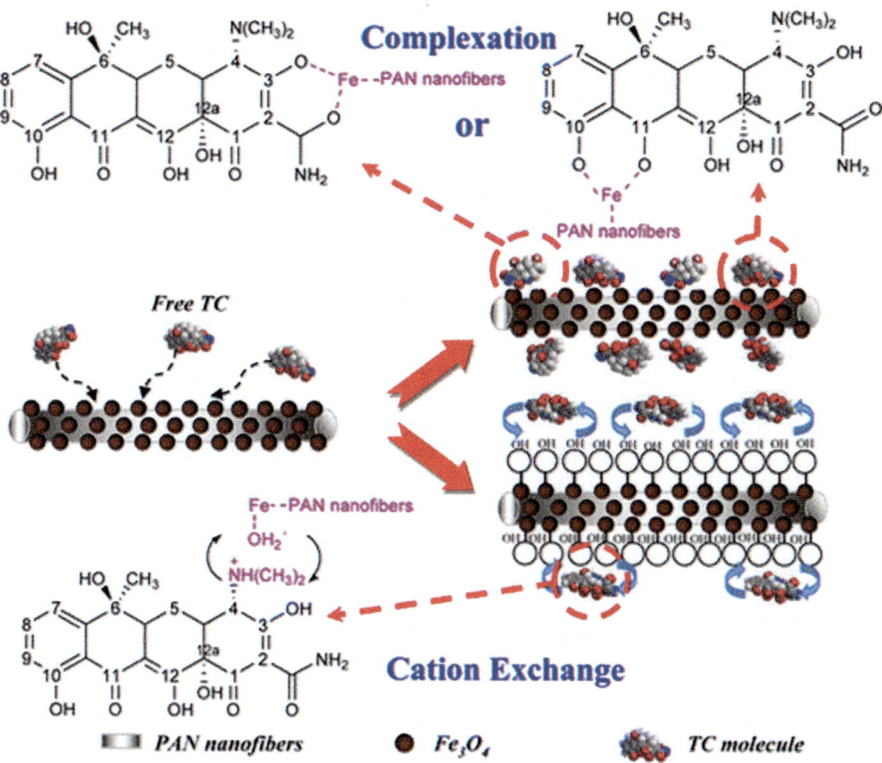

Fig. 6.26 Mechanistic illustration of adsorption of TC on Fe₃O₄/PAN composite NFs. Reproduced with permission from [101] © American Chemical Society

tion. The electrospinning researchers also paid significant attention toward the removal of pesticides. Lange et al. [103] reported the incorporation of Cu-BTC metal-organic framework (MOF-199) into polyacrylonitrile (PAN) nanofibers by the electrospinning process to remove methyl parathion, an OP pesticide in solution. Following 2 h of treatment, the PAN/MOF-199 membranes removed higher (88%) methyl parathion than the unmodified PAN membranes. Based on the outcome of the study, the functionalized nanofibrous membranes were suggested for their potential application in filtration media [103].

6.3 Conclusion and Future Prospects

The electrospun nanofibrous composite membrane has been recognized as an efficient filtering platform that provides competent ways for removing the pollutants from environment. This chapter summarized the recent advancements in the development of electrospun nanofibrous composite membranes as filtering media against various organic pollutants. The effective removal of organic pollutants from dyes to PAHs has been discussed comprehensively. However, the toxicity of functional materials and their environmental impact before and after coupling with nanofibers remain unclear and should be investigated systematically. Even though, there are studies focused on the leaching effect of functional nanomaterials from the nanofibers into filtering media, much concern should be paid in protecting functional molecules release from the nanofibrous filters to limit their impacts on environmental health.

References

1. Wen Y, Schoups G, van de Giesen N (2017) Organic pollution of rivers: combined threats of urbanization, livestock farming and global climate change. Sci Rep 7:43289
2. Petrie B, Barden R, Kasprzyk-Hordern B (2015) A review on emerging contaminants in wastewaters and the environment: current knowledge, understudied areas and recommendations for future monitoring. Water Res 72:3–27
3. Li QQ, Loganath A, Chong YS (2006) Persistent organic pollutants and adverse health effects in humans. J Toxicol Environ Health A 69:1987–2005
4. Aksu Z (2005) Application of biosorption for the removal of organic pollutants: a review. Process Biochem 40:997–1026
5. Ali I, Asim M, Khan TA (2012) Low cost adsorbents for the removal of organic pollutants from wastewater. J Environ Manag 113:170–183
6. Qu X, Alvarez PJJ, Li Q (2013) Applications of nanotechnology in water and wastewater treatment. Water Res 47:3931–3946
7. Zhang Y, Wu B, Xu H et al (2016) Nanomaterials-enabled water and wastewater treatment. NanoImpact 3–4:22–39
8. Bethi B, Sonawane SH, Bhanvase BA et al (2016) Nanomaterials-based advanced oxidation processes for wastewater treatment: a review. Chem Eng Process 109:178–189
9. Adeleye AS, Conway JR, Garner K et al (2016) Engineered nanomaterials for water treatment and remediation: costs, benefits, and applicability. Chem Eng J 286:640–662
10. Yoon K, Hsiao B, Chu B (2008) Functional nanofibers for environmental applications. J Mater Chem 18:5326–5334
11. Ramakrishna S, Fujihara K, Teo WE et al (2006) Electrospun nanofibers: solving global issue. Mater Today 9:40–50
12. Senthamizhan A, Balusamy B, Aytac Z et al (2016) Grain boundary engineering in electrospun ZnO nanostructures as promising photocatalyst. CrystEngComm 18:6341–6351
13. Kayaci F, Vempati S, Ozgit-Akgun C et al (2014) Enhanced photocatalytic activity of homoassembled zno nanostructures on electrospun polymeric nanofibres: a combination of atomic layer deposition and hydrothermal growth. Appl Catal B 156–157:173–183
14. Senthamizhan A, Balusamy B, Uyar T (2016) Glucose sensors based on electrospun nanofibers: a review. Anal Bioanal Chem 408:285–1306
15. Huang ZM, Zhang YZ, Kotaki M et al (2003) A review on polymer nanofibers by electrospinning and their applications in nanocomposites. Compos Sci Technol 63:2223–2253
16. Anitha S, Brabu B, Thiruvadigal DJ et al (2012) Preparation of free-standing electrospun composite ZnO membrane for antibacterial applications. Adv Sci Lett 5:468–474
17. Demirci S, Celebioglu A, Uyar T (2014) Surface modification of electrospun cellulose acetate nanofibers via RAFT polymerization for DNA adsorption. Carbohydr Polym 113:200–207
18. Senthamizhan A, Balusamy B, Celebioglu A et al (2016) "Nanotraps" in porous electrospun fibers for effective removal of lead (II) in water. J Mater Chem A 4:2484–2493

19. Wang P, Wang Y, Tong L (2013) Functionalized polymer nanofibers: a versatile platform for manipulating light at the nanoscale. Light Sci Appl 2:e102
20. Leung WH, Lo WH, Chan PH (2015) Amyloid fibrils as rapid and efficient nano-biosorbents for removal of dye pollutants. RSC Adv 5:90022–90030
21. Ahmad A, Mohd-Setapar SH, Chuong CS et al (2015) Recent advances in new generation dye removal technologies: novel search for approaches to reprocess wastewater. RSC Adv 5:30801–30818
22. Natarajan S, Bajaj HC, Tayade RJ (2017) Recent advances based on the synergetic effect of adsorption for removal of dyes from waste water using photocatalytic process. J Environ Sci. https://doi.org/10.1016/j.jes.2017.03.011
23. Dabrowski A (2001) Adsorption—from theory to practice. Adv Colloid Interf Sci 93:135–224
24. Chaúque EFC, Dlamini LN, Adelodun AA et al (2016) Electrospun polyacrylonitrile nanofibers functionalized with EDTA for adsorption of ionic dyes. Phys Chem Earth. https://doi.org/10.1016/j.pce.2016.10.008
25. Zarrini K, Rahimi AA, Alihosseini F et al (2017) Highly efficient dye adsorbent based on polyaniline-coated nylon-6 nanofibers. J Clean Prod 142:3645–3654
26. Bhaumik M, McCrindle R, Maity A (2013) Efficient removal of Congo red from aqueous solutions by adsorption onto interconnected polypyrrole–polyaniline nanofibers. Chem Eng J 228:506–515
27. Qureshi UA, Khatri Z, Ahmed F et al (2017) Electrospun zein nanofiber as green and recyclable adsorbent for the removal of reactive black 5 from aqueous phase. ACS Sustain Chem Eng. https://doi.org/10.1021/acssuschemeng.7b00402
28. Patel S, Hota G (2014) Adsorptive removal of malachite green dye by functionalized electrospun PAN nanofibers membrane. Fibers Polym 15:2272–2282
29. Miao YE, Wang R, Chen D et al (2012) Electrospun self-standing membrane of hierarchical SiO$_2$@γ-AlOOH(Boehmite) core/sheath fibers for water remediation. ACS Appl Mater Interfaces 4:5353–5359
30. Chen D, Liu C, Chen S et al (2016) Controlled synthesis of recyclable, porous FMO/C@TiO$_2$ core–shell nanofibers with high adsorption and photocatalysis properties for the efficient treatment of dye waste water. ChemPlusChem 81:282–291
31. Li S, Jia Z, Li Z (2016) Synthesis and characterization of mesoporous carbon nanofibers and its adsorption for dye in wastewater. Adv Powder Technol 27:591–598
32. Batool SS, Imran Z, Hassan S (2016) Enhanced adsorptive removal of toxic dyes using SiO$_2$ nanofibers. Solid State Sci 55:13–20
33. Im K, Nguyen DN, Kim S (2017) Graphene-embedded hydrogel nanofibers for detection and removal of aqueous-phase dyes. ACS Appl Mater Interfaces 9:10768–10776
34. Wang Y, Ding W, Jiao X et al (2014) Electrospun flexible self-standing silica/mesoporous alumina core–shell fibrous membranes as adsorbents toward Congo red. RSC Adv 4:30790–30797
35. Patel S, Hota G (2016) Iron oxide nanoparticle-immobilized PAN nanofibers: synthesis and adsorption studies. RSC Adv 6:15402–15414
36. Uyar T, Havelund R, Nur Y et al (2009) Molecular filters based on cyclodextrin functionalized electrospun fibers. J Membrane Sci 332:129–137
37. Uyar T, Havelund R, Hacaloglu J et al (2010) Functional electrospun polystyrene nanofibers incorporating alpha, beta and gamma cyclodextrins: comparison of molecular filter performance. ACS Nano 4:5121–5130
38. Egede EJ, Jones H, Cook B et al (2016) Application of microalgae and fungal-microalgal associations for wastewater treatment. In: Purchase D (ed) Fungal applications in sustainable environmental biotechnology, springer international publishing, Switzerland, pp 143–181
39. Sarioglu OF, San Keskin NO, Celebioglu A et al (2017) Bacteria encapsulated electrospun nanofibrous webs for remediation of methylene blue dye in water. Colloids Surf B Biointerfaces 152:245–251
40. Salalha W, Kuhn J, Dror Y et al (2006) Encapsulation of bacteria and viruses in electrospun nanofibers. Nanotechnology 17:4675–4681

41. Maurya NS, Mittal AK, Cornel P et al (2006) Biosorption of dyes using dead macro fungi: effect of dye structure: ionic strength and pH. Bioresour Technol 97:512–521
42. Sarioglu OF, Celebioglu A, Tekinay T et al (2015) Evaluation of contact time and fiber morphology on bacterial immobilization for development of novel surfactant degrading nanofibrous webs. RSC Adv 5:102750–102758
43. San NO, Celebioglu A, Tumtas Y et al (2014) Reusable bacteria immobilized electrospun nanofibrous webs for decolorization of methylene blue dye in wastewater treatment. RSC Adv 4:32249–32255
44. San Keskin NO, Celebioglu A, Uyar T et al (2015) Microalgae immobilized by nanofibrous web for removal of reactive dyes from wastewater. Ind Eng Chem Res 54:5802–5809
45. Kim KH, Jahan SA, Kabir E (2013) A review on human health perspective of air pollution with respect to allergies and asthma. Environ Int 59:41–52
46. De Crom J, Claeys S, Godayol A et al (2010) Sorbent-packed needle microextraction trap for benzene, toluene, ethylbenzene, and xylenes determination in aqueous samples. J Sep Sci 33:2833–2840
47. Atkinson R (2000) Atmospheric chemistry of VOCs and NOx. Atmos Environ 34:2063–2101
48. Konieczny K, Bodzek M, Panek D (2008) Removal of volatile compounds from the wastewaters by use of pervaporation. Desalination 223:344–348
49. Son YS (2017) Decomposition of VOCs and odorous compounds by radiolysis: a critical review. Chem Eng J 316:609–622
50. Hirota K, Sakai H, Washio M et al (2004) Application of electron beams for the treatment of VOC streams. Ind Eng Chem Res 43:1185–1191
51. Hakim M, Broza YY, Barash O et al (2012) Volatile organic compounds of lung cancer and possible biochemical pathways. Chem Rev 112:5949–5966
52. Delfino RJ, Gong H, Linn WS et al (2003) Respiratory symptoms and peak expiratory flow in children with asthma in relation to volatile organic compounds in exhaled breath and ambient air. J Expo Anal Environ Epidemiol 13:348–363
53. Gałęzowska G, Chraniuk M, Wolska L (2016) In vitro assays as a tool for determination of VOCs toxic effect on respiratory system: a critical review. TrAC Trends Anal Chem 77:14–22
54. Al-Dawery S (2013) Methanol removal from methanol-water mixture using municipal activated sludge. J Eng Sci Technol 8:578–587
55. Aliabadi M, Aroujalian A, Raisi A (2012) Removal of styrene from petrochemical wastewater using pervaporation process. Desalination 284:116–121
56. Kujawa J, Cerneaux S, Kujawski W (2015) Highly hydrophobic ceramic membranes applied to the removal of volatile organic compounds in pervaporation. Chem Eng J 260:43–54
57. Son YS, Kim P, Park JH et al (2013) Decomposition of trimethylamine by an electron beam. Plasma Chem Plasma Process 33:1099–1109
58. Vane LM, Alvarez FR (2002) Full-scale vibrating pervaporation membrane unit: VOC removal from water and surfactant solutions. J Membr Sci 202:177–193
59. Delimaris D, Ioannides T (2008) VOC oxidation over $MnOx-CeO_2$ catalysts prepared by a combustion method. Appl Catal B Environ 84:303–312
60. Balamurugan R, Sundarrajan S, Ramakrishna S (2011) Recent trends in nanofibrous membranes and their suitability for air and water filtrations. Membranes 1:232–248
61. Thavasi V, Singh G, Ramakrishna S (2008) Electrospun nanofibers in energy and environmental applications. Energy Environ Sci 1:205–221
62. Mirjalili M, Zohoori S (2016) Review for application of electrospinning and electrospun nanofibers technology in textile industry. J Nanostruct Chem 6:207–213
63. Zhu M, Han J, Wang F et al (2017) Electrospun nanofibers membranes for effective air filtration. Macromol Mater Eng 302:1600353
64. Haider A, Haider S, Kang IK (2015) A comprehensive review summarizing the effect of electrospinning parameters and potential applications of nanofibers in biomedical and biotechnology. Arab J Chem. https://doi.org/10.1016/j.arabjc.2015.11.015
65. Sundarrajan S, Tan KL, Lim SH et al (2014) Electrospun nanofibers for air filtration applications. Procedia Eng 75:159–163

66. Sahay R, Kumar PS, Sridhar R et al (2012) Electrospun composite nanofibers and their multifaceted applications. J Mater Chem 22:12953–12971
67. Chu L, Deng S, Zhao R et al (2015) Adsorption/desorption performance of volatile organic compounds on electrospun nanofibers. RSC Adv 5:102625–102632
68. Shim WG, Kim C, Lee JW et al (2006) Adsorption characteristics of benzene on electrospun-derived porous carbon nanofibers. J Appl Polym Sci 102:2454–2462
69. Bai Y, Huang ZH, Wang MX et al (2013) Adsorption of benzene and ethanol on activated carbon nanofibers prepared by electrospinning. Adsorption 19:1035–1043
70. Scholten E, Bromberg L, Rutledge GC et al (2011) Electrospun polyurethane fibers for absorption of volatile organic compounds from air. ACS Appl Mater Interfaces 3:3902–3909
71. Feng C, Khulbe KC, Tabe S (2012) Volatile organic compound removal by membrane gas stripping using electro-spun nanofiber membrane. Desalination 287:98–102
72. Guo Z, Huang J, Xue Z et al (2016) Electrospun graphene oxide/carbon composite nanofibers with well-developed mesoporous structure and their adsorption performance for benzene and butanone. Chem Eng J 306:99–106
73. Uyar T, Havelund R, Nur Y et al (2010) Cyclodextrin functionalized poly(methyl methacrylate) (PMMA) electrospun nanofibers for organic vapors waste treatment. J Membr Sci 365:409–417
74. Kayaci F, Uyar T (2014) Electrospun polyester/cyclodextrin nanofibers for entrapment of volatile organic compounds. Polym Eng Sci 54:2970–2978
75. Kayaci F, Sen HS, Durgun E et al (2015) Electrospun nylon 6,6 nanofibers functionalized with cyclodextrins for removal of toluene vapor. J Appl Polym Sci 132:41941
76. Celebioglu A, Uyar T (2013) Electrospun gamma-cyclodextrin (γ-CD) nanofibers for the entrapment of volatile organic compounds. RSC Adv 3:22891–22895
77. Celebioglu A, Sen HS, Durgun E et al (2016) Molecular entrapment of volatile organic compounds (VOCs) by electrospun cyclodextrin nanofibers. Chemosphere 144:736–744
78. Kim HJ, Pant HR, Choi NJ et al (2013) Composite electrospun fly ash/polyurethane fibers for absorption of volatile organic compounds from air. Chem Eng J 230:244–250
79. Ge JC, Kim JH, Choi NJ (2016) Electrospun polyurethane/loess powder hybrids and their absorption of volatile organic compounds. Adv Mater Sci Eng 2016:8521259
80. Ge JC, Choi N (2017) Fabrication of functional polyurethane/rare earth nanocomposite membranes by electrospinning and its VOCs absorption capacity from air. Nano 7:60
81. Haritash AK, Kaushik CP (2009) Biodegradation aspects of polycyclic aromatic hydrocarbons (PAHs): a review. J Hazard Mater 169:1–15
82. Kaushik CP, Haritash AK (2006) Polycyclic aromatic hydrocarbons (PAHs) and environmental health. Our Earth 3:1–7
83. Samanta SK, Singh OV, Jain RK (2002) Polycyclic aromatic hydrocarbons: environmental pollution and bioremediation. Trends Biotechnol 20:243–248
84. Kim KH, Jahan SA, Kabir E et al (2013) A review of airborne polycyclic aromatic hydrocarbons (PAHs) and their human health effects. Environ Int 60:71–80
85. Bansal V, Kim KH (2015) Review of PAH contamination in food products and their health hazards. Environ Int 84:26–38
86. Jarvis IW, Dreij K, Mattsson Å et al (2014) Interactions between polycyclic aromatic hydrocarbons in complex mixtures and implications for cancer risk assessment. Toxicology 321:27–39
87. Rubio-Clemente A, Torres-Palma RA, Peñuela GA (2014) Removal of polycyclic aromatic hydrocarbons in aqueous environment by chemical treatments: a review. Sci Total Environ 478:201–225
88. Abdel-Shafy HI, Mansour MSM (2016) A review on polycyclic aromatic hydrocarbons: source, environmental impact, effect on human health and remediation. Egypt J Pet 25:107–123
89. Paszkiewicz M, Caban M, Bielicka-Giełdoń A et al (2017) Optimization of a procedure for the simultaneous extraction of polycyclic aromatic hydrocarbons and metal ions by functionalized and non-functionalized carbon nanotubes as effective sorbents. Talanta 165:405–411

90. Li J, Zhou QX, Liu YL et al (2017) Recyclable nanoscale zero-valent iron-based magnetic polydopamine coated nanomaterials for the adsorption and removal of phenanthrene and anthracene. Sci Technol Adv Mater 18:3–16

91. Perreault F, Fonseca de Faria A, Elimelech M (2015) Environmental applications of graphene-based nanomaterials. Chem Soc Rev 44:5861–5896

92. Qi D, Kang X, Chen L et al (2008) Electrospun polymer nanofibers as a solid-phase extraction sorbent for the determination of trace pollutants in environmental water. Anal Bioanal Chem 390:929–938

93. Dai Y, Niu J, Yin L et al (2011) Sorption of polycyclic aromatic hydrocarbons on electrospun nanofibrous membranes: sorption kinetics and mechanism. J Hazard Mater 192:1409–1417

94. Dai Y, Yin L, Niu J (2011) Laccase-carrying electrospun fibrous membranes for adsorption and degradation of PAHs in shoal soils. Environ Sci Technol 45:10611–10618

95. Dai Y, Niu J, Yin L et al (2013) Laccase-carrying electrospun fibrous membrane for the removal of polycyclic aromatic hydrocarbons from contaminated water. Sep Purif Technol 104:1–8

96. Kayaci F, Aytac Z, Uyar T (2013) Surface modification of electrospun polyester nanofibers with cyclodextrin polymer for the removal of phenanthrene from aqueous solution. J Hazard Mater 261:286–294

97. Celebioglu A, Demirci S, Uyar T (2014) Cyclodextrin-grafted electrospun cellulose acetate nanofibers via "click" reaction for removal of phenanthrene. Appl Surf Sci 305:581–588

98. Sui Q, Cao X, Lu S et al (2015) Occurrence, sources and fate of pharmaceuticals and personal care products in the groundwater: a review. Emerg Contam 1:14–24

99. Hao R, Xiao X, Zuo X et al (2012) Efficient adsorption and visible-light photocatalytic degradation of tetracycline hydrochloride using mesoporous BiOI microspheres. J Hazard Mater 209–210:137–145

100. Le-Minh N, Khan SJ, Drewes JE et al (2010) Fate of antibiotics during municipal water recycling treatment processes. Water Res 44:4295–4323

101. Liu Q, Zhong LB, Zhao QB et al (2015) Synthesis of Fe₃O₄/Polyacrylonitrile composite electrospun nanofiber mat for effective adsorption of tetracycline. ACS Appl Mater Interfaces 7:14573–14583

102. Banks KE, Hunter DH, Wachal DJ (2005) Chlorpyrifos in surface waters before and after a federally mandated ban. Environ Int 31:351–356

103. Lange LE, Ochanda FO, Obendorf SK et al (2014) CuBTC metal-organic frameworks enmeshed in polyacrylonitrile fibrous membrane remove methyl parathion from solutions. Fibers Polym 15:200–207

Chapter 7
Electrospun Filters for Oil–Water Separation

Mohammad Mahdi A. Shirazi and Morteza Asghari

Abstract The separation of oil from water in emulsions is a challenging and costly problem in several industrial sectors. Oil–water separation using electrospun fibers is a relatively new but highly promising technique. Highly specific surface areas, interconnected pore structures, and the potential to incorporate active chemistry on a nanoscale surface have made the electrospun fibers a promising versatile platform for the treatment of oily wastewaters. In this chapter, we summarize the applications and recent developments of electrospun filters and membranes for oil–water separation. First, conventional processes and materials for oily wastewater treatment are summarized. Then, the electrospinning technique, important operating parameters, properties of electrospun fibers for oil–water separation, and filtration mechanisms are introduced. Afterward, applications of electrospun fibers for oil–water separation, including coalescing filtration and membrane separation for oily wastewater treatment, are comprehensively discussed. Finally, the challenges and perspectives for the future of this subject are discussed.

7.1 Introduction

Recently, the increasing world population, urbanization, and industrialization have led to increase in the global energy demand. This has consequently led to generate large amounts of highly polluted wastewaters around the world [1]. Oily wastewater is among the most challenging ones, usually generated in different industries including oil production and processing [2, 3], petrochemical plants and refineries [4, 5], food industry and biotechnology [6, 7], etc. Moreover, oil leakage and oil spills

M. M. A. Shirazi
Separation Processes Research Group (SPRG), Department of Chemical Engineering,
University of Kashan, Kashan, Iran

Membrane Industry Development Institute, Tehran, Iran

M. Asghari (✉)
Separation Processes Research Group (SPRG), Department of Chemical Engineering,
University of Kashan, Kashan, Iran
e-mail: asghari@kashanu.ac.ir

© Springer International Publishing AG, part of Springer Nature 2018
M. L. Focarete et al. (eds.), *Filtering Media by Electrospinning*,
https://doi.org/10.1007/978-3-319-78163-1_7

often occur during oil production, transportation, and storage which is another critical environmental concern [8, 9]. The most important concerns of oily wastewater release to the environment include affecting drinking water sources (surface water and groundwater sources), endangering aquatic sources, atmospheric pollution, affecting crop production, and destructing natural landscape. All these issues can directly affect the human health [10, 11]. Moreover, reclaiming the water and reclaiming the precious oil resource are other important issues.

All mentioned concerns could be more highlighted by investigating the strict standards for discharging oily wastewaters. The permitted oil limits for treated wastewater discharge to the environment in Australia and the USA (daily average) are 30 mg/L and 42 mg/L, respectively. Some other countries have implemented more stringent standards for discharging oily wastewater. For instance, the monthly average limit of oil and grease prescribed by the People Republic of China is 10 mg/L [2].

Furthermore, due to the large amount of oily wastewater being generated, many countries with oil/gas fields (i.e., underdeveloping countries) and also developed countries with industrial sectors are increasingly focusing on efforts to find efficient and cost-effective treatment approaches for oily wastewater management (oil–water separation). It is worth quoting that the treated wastewater has potential uses as an alternative source for irrigation, livestock, or wildlife watering, and also for a number of industrial uses such as dust control, vehicle washing, power-plant makeup, and most importantly for fire control [2, 4].

All abovementioned concerns reflect in the fact that developing advanced materials to effectively and eco-friendly separate the oil–water mixtures (oily wastewater treatment) has become urgent and needs significant research efforts.

7.2 Oil–Water Separation Processes and Materials

7.2.1 Conventional Processes for Oil–Water Separation

Oily wastewater treatment has been mainly conducted by a number of physical, chemical, and biological methods including adsorption [12], biological treatment [13], and electro-/coagulation [14]. Table 7.1 summarizes an overview of conventional methods for oily wastewater treatment. However, the proposed conventional methods do not provide a completely satisfactory solution. In one hand, chemical methods are fast, efficient, and mature ones, but the use of toxic/harmful chemicals and the resulting secondary effluents are their major drawbacks. The same disadvantages are found for biological methods, in addition to low overall process rate. On the other hand, physical methods, such as adsorption, need external adsorbents which present the disadvantages of low adsorption rate, small separation capacity, and restricted recyclability. Further information of conventional treatment methods and their *Pros.* and *Cons.* could be found in the literature [2, 4, 15–18].

Table 7.1 Overview of the conventional methods for oily wastewater treatment

Separation process	Techniques	Advantage	Disadvantage
Chemical treatment	• Chemical precipitation • Chemical oxidation • Electrochemical treatment • Photocatalytic treatment • Fenton process • Ozone treatment • Demulsifier • Room temperature ionic liquids	• Well-developed • Fast treatment • Applicable for a wide range of treatments	• Using chemical additives • Causing secondary pollution • Costly equipment • Corrosion problem
Physical treatment	• Sand filters • Adsorption • Cyclones • Evaporation • Dissolved air floatation/ precipitation • C-tour • Freeze–thaw/ evaporation • Electrodialysis	• Well-developed • Applicable for a wide range of treatments	• Low efficiency • High maintenance costs • Corrosion
Biological treatment	• Activated sludge • Trickling filter • Sequencing batch reactor • Chemostate reactor • Biological oxidation • Biological aerated filters • Lagoons	• Applicable for a wide range of treatments	• Slow treatment process • Causing secondary effluent • Limitations regarding type of used microorganisms • Underdeveloped

7.2.2 Conventional Filtration and Membrane Processes for Oil–Water Separation

Filtration and membrane separation processes have been investigated to be one of the most efficient methods for treatment of oily wastewater streams [19]. In comparison to the conventional treatment methods (see Table 7.1), filters and membranes offer higher oil removal efficiency, more stable effluent quality, and lower costs and energy consumptions [20].

Polymeric filters and membranes can be fabricated using different methods, including phase inversion, interfacial polymerization, stretching, track-etching [21], and recently electrospinning [22]. The selection of fabrication technique depends on a number of items such as polymer–solvent type and composition, desired structure, and the target application [23]. Each technique has its own benefits and limitations as briefly summarized in Table 7.2. Detailed information regarding the conventional filters and membranes for oil–water separation could be found elsewhere [17, 22, 24–28].

Table 7.2 An overview of membrane fabrication techniques

Fabrication	Description	Separation	Pore size	Challenges
Phase inversion	It's a demixing procedure where the initially homogeneous polymer solution is transformed in a controlled manner from a liquid phase to a solid state	RO UF	RO: 2–5 Å UF: 0.01– 0.1 μm	• Limited range of pore size and its distribution • Residual solvent content
Interfacial polymerization	This is one of the most prominent fabrication technique for commercial TFC membranes, mostly has developed for NF membranes	NF	0.001– 0.01 μm	• Controlling the selective layer thickness • Residual solvent content
Stretching	Polymer is heated (above its melting point) and is converted into thin film, then followed by stretching to make it porous This is a solvent free fabrication	Filtration, MF and MD	0.1–1 μm	• Controlling the pore size • Lack of mechanical strength
Track-etching	Irradiating a nonporous polymeric film energetic heavy ions	Filtration, MF and MD	0.1– 10 μm	• Controlling the pore size • Lack of mechanical strength for high porosity values • Costly

RO reverse osmosis, *NF* nanofiltration, *UF* ultrafiltration, *MF* microfiltration, *MD* membrane distillation

However, one of the promising alternatives for oil–water separation is developing materials with special wettable characteristics that have distinct opposite affinities towards oil and water. In this regard, two main types of special wettable materials have been developed, i.e., hydrophobic and oleophilic materials, and hydrophilic and oleophobic materials [22, 29]. The wetting behavior of materials is a function of their surface chemistry and it can be further boosted by material's surface architectures [30, 31]. In other words, when the surface tension of a material lies between those of water and oil, it performs as a hydrophobic (oleophilic) one [32]. In the next section, special wettable materials for oil–water separation are briefly introduced and discussed.

7.2.3 Conventional Wettable Materials for Oil–Water Separation

The developed special wettable materials for oily wastewater treatment (oil–water separation) are classified into two major types based on the method used for separation, including filtration materials (meshes, textiles, filters, and membranes) and adsorption materials (powders and 3D networks/webs) (see Fig. 7.1).

Fig. 7.1 A general scheme of the materials for oil–water separation developed so far

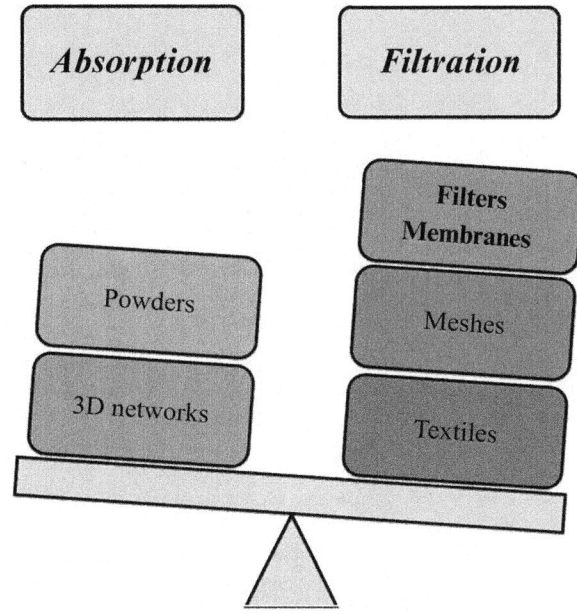

A filter/membrane allows either water or oil to permeate, while preventing the other one from passing through [33]. Adsorptive material selectively absorbs either water or oil onto its surface and into its interior empty voids but repels the other phases when being exposed to an oily wastewater stream (oil–water emulsion) [34]. Each one of these treatment methods has its own advantages and disadvantages. However, in this chapter, only oil–water separation filters and membranes are discussed and further information regarding adsorptive materials for oil–water separation can be found elsewhere [29, 35].

Some of the most extensively studied filters for oily wastewater treatment are metallic meshes. This is due to their excellent mechanical strength, industrial scale production, and high filtration fluxes. Moreover, reactive metal surface provides a wide range of possible surface modifications. So far, copper and stainless steel are the most investigated meshes for oil–water separation purposes. Oil–water separation meshes can be categorized into four major types, i.e., hydrophobic–oleophilic meshes [20], hydrophilic–oleophobic meshes [36], superhydrophilic and underwater superoleophobic [37], and responsive/switchable wetting meshes [38]. Table 7.3 listed some metallic meshes for oil–water separation.

Although metallic meshes with special wettable characteristics have shown promising performance for selective oil–water separation, there are some weak points, including high cost for the used substrate, corrosion instability, heavy mass, and disposal/reuse concerns. To solve these concerns, textiles and fabrics which are cheaper, lighter, more flexible, and more corrosion resistive than metallic meshes have been investigated [46, 47]. However, some drawbacks, such as low separation capacity on one hand, and increasing the environmental awareness and tighter

Table 7.3 An overview of the oil–water separation meshes

Metallic mesh	Description	Substrate and coating	Surface contact angle with water/oil	Reference
Hydrophobic and oleophilic	• Oil has a lower surface energy than water • So, separation meshes coated with hydrophobic agents are the most straightforward approach	Stainless steel and PTFE	156°	[39]
		Cu mesh and polyurethane	165.1°	[40]
		Stainless steel and epoxy/attapulgite	160°	[41]
		Stainless steel and CNT	145-150°	[42]
Hydrophilic and oleophobic	• These materials can selectively let water pass through • Theoretically, it is hard to prepare such materials, due to higher surface tension of water than oil	Stainless steel and PDDA–PFO	157°	[37]
Superhydrophilic and underwater superoleophobic	• Theoretically, a surface with superhydrophilicity characteristic in air shows superoleophobicity feature underwater • This is due to water molecules trapped onto material's hierarchical surface structure to reach a low energy state	Cu mesh and $Cu(OH)_2-(NH_4)_2S_2O_8$	>150°	[43]
		Cu mesh and $Cu(OH)_2 - 0.05$ m $K_2S_2O_8/$ 1.0 m NaOH	166.2°	[44]
Responsive/ Switchable	• These materials can change their wettability upon using an external stimulus or treatment with certain chemicals • Materials with responsive wettability can switch from wetting to antiwetting, or vice versa • As a result, they can be water removing or oil removing, depending on the conditions	A pH responsive copper mesh. Surface modification with a thiol mixture of $HS(CH_2)_9CH_3$ and $HS(CH_2)_{10}COOH$		[45]

PTFE polytetrafluoroethylene, *CNT* carbon nanotube, *PDDA* poly(diallyldimethylammonium chloride), *PFO* sodium perfluorooctanoate regulations on another hand, have led to put forward high-tech strategies to treat the oily wastewaters. Further information regarding the conventional materials, e.g., meshes and fabrics/textiles can be found in the literature [20, 29, 48].

7.3 Electrospun Filters and Membranes

Recently, electrospinning (ES) method has been progressively more investigated for the fabrication of microporous filters and membranes [49]. The worldwide attention toward electrospinning method for fabricating polymeric filters is due to its simple and versatile nature to create fibers in the range of micro- to nanosize through an electrically charged jet of a polymer solution or melt [50]. Thus, the development of electrospun filters and membranes with a nanofibrous three-dimensional (3D) interconnected pore structure has opened up a new window for oil–water filtration during recent years [22, 50].

7.3.1 Electrospinning Technique

Electrospinning is a technique that can generate fibers with diameter as low as ~10 nm [51] from any conductive viscoelastic liquid which can solidify [52]. Further to versatility in selecting the raw material, electrospinning is a simple technique that has a number of options to control fibers' diameter, morphology, surface topology, and also fibers' arrangement.

Electrospinning is a versatile technique for fabricating nanofibrous membranes for a wide range of applications including desalination, water/wastewater treatment, and oil–water separation [35, 50, 53]. This is due to various advantages of this technique over other conventional methods for fabricating polymeric membranes (see Table 7.2) including:

- *High surface area to volume ratio*: the micro- and/or nanodimension of as-spun nanofibers provides a high surface area to volume ratio. This is an attractive feature, specifically for membrane applications.
- *Variety of polymers and solvents utilized*: this technique has been used to fabricate nanofibers from a wide range of polymer–solvent mixtures.
- *Low start-up costs*: this technique is generally cheaper than some other fabricating techniques.
- *Variety of substrates utilized*: electrospun fibers can be deposited on surface of various substrates.
- *Functionalization*: electrospun nanofibers can be easily functionalized to improve the separation performance.
- *Commercial applications*: air/dust filters, face mask, and liquid filters/membranes are some examples of commercialized products.

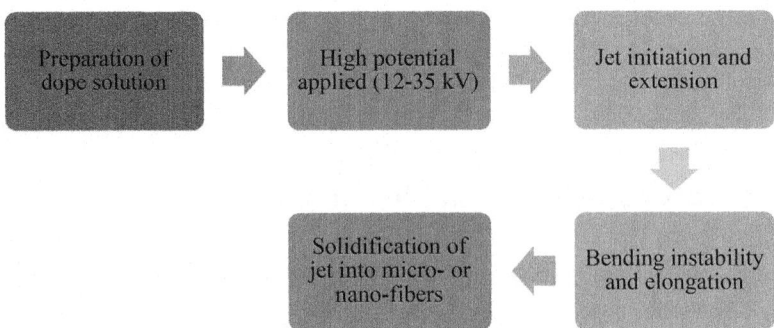

Fig. 7.2 Five main steps to fabricate electrospun fibers

7.3.2 Processing Parameters

Figure 7.2 shows the five main steps of the electrospinning process, which start from a solution preparation (polymer–solvent mixture), followed by applying a high voltage (between a needle and a metal collector), and end with solidification of the jet into micro- or nanosized fibers.

It is worth quoting that the final properties of electrospun fibers greatly depend on the processing parameters, which can be classified into three major groups, including solution parameters (i.e., polymer–solvent ratio, viscosity, surface tension, conductivity, dielectric constant, and solvent volatility), process parameters (i.e., electrostatic potential, electric field strength, electrostatic field shape, tip-to-collector distance, solution flow rate, and needle diameter), and environment parameters (i.e., temperature, humidity, pressure, and atmospheric composition) [53]. A number of features such as surface topography, fiber morphology, and orientation are largely dictated by solution properties and operating conditions [54]. Since the rheology of the polymer solution is vital to the electrospun fiber formation, polymer molecular weight and concentration directly affect fiber properties [55]. Conductivity of the polymer solution is also known to change properties of the resultant fibers [56]. Moreover, electrospun fiber properties are directly affected by process parameters and ambient conditions, as mentioned above. Further detailed information on the effect of operating parameters on the electrospinning process can be found in the literature [22].

7.3.3 Properties of Electrospun Fibers for Oil–Water Filtration

Like oil–water separation meshes, electrospun fibers can be categorized based on their hydrophobicity features. Hydrophobicity is one of the most important characteristics of electrospun filters for oil–water separation. It is worth quoting that the term *Hydrophobicity* is based on the physical property of the surface that repels water [52]. When the surface water contact angle of a material is greater than 90°, the material is considered to be hydrophobic. Superhydrophobic surface, however, shows a water contact angle of >150° [57]. It should be noted that a hydrophobic filter can be developed via the electrospinning of low surface energy materials or through surface modification of as-spun filters. However, for preparing a superhydrophobic filter not only the low surface energy material but also the hierarchical surface structure is required [52].

Surface morphology- (smooth, rough, or a combination of these textures) is another important characteristic of electrospun filters/membranes [58]. Rough surface can enhance the oil adhesion to the fibers' surface [59]. Moreover, higher surface roughness can increase the hydrophobicity of the filter [60]. It is reported in the literature that the high porosity is another key factor to obtain high filtration flux [23]. One of the most important advantages of the electrospun filters and membranes is the higher porosity compared to the other conventional membranes. This is due to the 3D interconnected structure of electrospun membranes. For instance, Mirtalebi et al. [61] studied the comparative performance of commercial and electrospun nylon membranes for microfiltration of petrochemical wastewater containing coke particles. Results indicated that all used membranes showed the same coke rejection of 99%. However, considerably higher permeate flux was achieved for the electrospun membrane due to its higher porosity.

Further to the abovementioned features, fibers' diameter, mechanical strength, and chemical stability are the other important parameters that can affect the electrospun filters' performance [62].

7.3.4 Filtration Mechanisms

It is worth noting that the commercial polymeric filters/membranes, such as those made of PTFE, polyethersulfone (PES), PVDF, CA, mixed cellulose ester (MCE), etc., have two-dimensional (2D) composite structures, consisting of a thin microporous active (selective) layer which is either casted or coated on top of a porous support layer. The main separation mechanism for membranes with this 2D structure is almost sieving [63, 64].

In contrast to the 2D structure of commercial membranes, the separation mechanism of the electrospun membranes with 3D interconnected structure is a combination of sieving and depth filtration. In fact, randomly structured micro- and nanosized fibers lead to fabrication of a highly porous membrane with high selective separation capacity. The multi-objective separation mechanism of electrospun filters/membranes (screening, depth filtration, and adsorption) could be the cause of their superior separation performance. These features are obviously interesting for the filtration of oily wastewater streams [33, 65].

7.4   Applications of Electrospun Filters for Oil–Water Separation

Special wettable materials, i.e., metallic meshes and fabrics (textiles), as introduced earlier, have demonstrated their attractive functions in oil–water separation. However, it should be noted that these materials are mostly applied to separate layered oil–water mixtures and they cannot effectively separate oil–water emulsions. This is due to their large pore sizes which are usually larger than the emulsified oil/water droplets.

Recently, much attention has been paid to electrospun nanofibers for oil–water separation, due to their smaller pore size, higher surface area, highly porous structure, and remarkable separation efficiency. However, to ensure an efficient oil–water separation process, the selection of a suitable separation barrier (filter or membrane) is essential. In this regard, electrospun filters for the proposed treatment, i.e., oil–water separation, can be categorized into two main groups: coalescing filters and membranes. For both filters and membranes, there are some key parameters related to their performance. These parameters include, but are not limit to, pore size and distribution, porosity, surface energy, chemical compatibility, and surface morphological and topographical features.

7.4.1   Electrospun Filters: Coalescing Filtration

When two liquids are immiscible, such as oil and water, they can form an emulsion [66]. In the resultant mixture, the dispersed liquid forms droplets in the continuous phase, i.e., oil in water or vice versa [67]. Emulsions can be classified into primary and secondary emulsions. Dispersed droplets in the former type typically have diameters bigger than 100 μm, while in the secondary emulsions, the droplet diameter is typically less than 100 μm [68]. Both the emulsion types are serious challenges for different industries, specifically the petrochemical industry, due to their stability and difficulty of separation through the conventional methods (see Table 7.1), especially those of the second type.

Liquid–liquid coalescers are used to accelerate the merging of fine oil/water droplets to form a less number of droplets, but with a greater diameter. Settling of the larger droplets downstream of the coalescer system then requires considerably less residence time. As a result, this can be translated to less steel requirement, smaller equipment size, and faster and more cost-effective treatment process. Coalescers exhibit a three-step method of operation including: (1) collection of individual droplets, (2) combining of several small droplets into larger ones, and (3) rise of the enlarged droplets by gravity force to make an oil layer [69, 70].

Therefore, coalescing filtration performance depends on various parameters such as feed flowrate, oil droplets' size and distribution, filter bed depth, and surface characteristics of the filter material [71]. Fibrous porous filters provide the advantage of high filtration efficiency at reasonable costs and energy consumption. Conventionally, micro-glass fibers are most frequently used for oil–water coalescing filtration [72]. However, due to submicron size of the glass fibers, these filters cannot coalesce the oil droplets with size below 50 μm [73].

Recently, electrospun nanofibers have been used to enhance the oil–water separation of coalescing filtration systems. Table 7.4 reports some literature examples of electrospun nanofibrous filters for oil–water coalescing filtration. As could be observed, most of the researches on coalescing filters used the electrospun polystyrene nanofibers to improve the oil–water separation efficiency. This is due to its excellent characteristics, such as remarkable hydrophobicity, and good performance for the proposed application [80]. Polystyrene filters can easily separate liquids with low surface tension, such as mineral oil, edible oil, gasoline, and diesel from water. These filters have shown the ability of oil–water separation, even in a single step of only few-minutes' filtration. Moreover, excellent hydrophobicity even after many cycles of oil–water filtration is another interesting feature of polystyrene. These facts can confirm applicability of polystyrene filters for removal of many types of organic solvents and oil in industrial scale.

Experiments have shown that the addition of small amounts of polystyrene nanofibers significantly enhances the coalescing efficiency of filters [74]. However, it should be noted that this can increase the filter's pressure drop. So, there exists an optimum amount of nanofibers to be added to the coalescing filtration media for oil–water separation [75]. This can balance the desired improvement in coalescence efficiency and the undesired increase in the pressure drop [81].

7.4.2 Electrospun Membranes

In coalescing filtration, as mentioned, both ingredients of oil–water emulsion pass through the filter. Therefore, the role of the filter is to collect and enlarge fine oil drops to facilitate separation [81]. On the other hand, in membrane filtration, only one of the phases, depending on the hydrophobic or oleophobic characteristics of the membrane, can pass through the pores (permeate stream) and the other one remains in the retentate stream [17]. Most of the used conventional membranes with 2D structure for oil–water separation are applying the hydrostatic pressure difference as the driving force, i.e., pressure-driven membrane processes (see Table 7.5). In a same way, electrospun oil–water separation membranes can be classified as pressure-driven membranes, however, with lower operating pressure (0.1–0.3 bar). It is worth quoting that most of these nanofibrous membranes, based on the pore size (0.1–5 μm), are in the range of microfiltration process [85]. In other words, the large pore size of an electrospun membrane makes it unsuitable for filtration purposes below the microfiltration range, such as ultrafiltration. However, in ultrafiltration and nanofiltration processes, mostly for water/wastewater treatment, the electrospun membranes are often used as the support layer for the selective barrier [86].

Table 7.4 Summary of research on coalescing filtration of oil-in-water streams using electrospun filters

Polymer–solvent dope	Electrospinning	Fibers characteristics	Separation	Reference
Polymer: polystyrene Mw: 190,000 Solvent: DMAc	• High voltage: 18 kV • Needle tip diameter: 18 gauge • Tip-to-collector distance: 17 cm • Dope flowrate: 0.1 ml/min • Electrospinning time: 30 min • Ambient: $T = 25\ °C$, humidity = 20–25% *Post-treatment*: thermal treatment through contact heating	*Before thermal treatment* • Mean fiber diameter: 0.412 μm • Roughness: 290 nm • Contact angle: 135° *After thermal treatment* • Mean fiber diameter: 0.491–1.456 μm • Roughness: 230 nm • Contact angle: 156°	Oily wastewater treatment	[74]
Polymer: poly(meta-phenylene isophthalamide) Solvent: DMAc	• High voltage: 15 kV • Needle tip diameter: 1 mm • Tip-to-collector distance: 20 cm • Dope flowrate: 3.6 μL/min	*Fibers mean diameter* • 25 μm and 150 nm *Filters porosity* • 91.6–92.3%	Oil-in water separation Separation efficiency improved from 71% to 84%	[75]
Polymer: expanded polystyrene Solvent: DMAc	• High voltage: 20 kV • Needle: a glass pipette • Tip diameter: 1 mm • Tip-to-collector distance: 20 cm	*Filter mass* • 0.81–0.92 g *Separation efficiency* • 64–91%	Oil–water separation	[76]
Polymer: recycled expanded polystyrene Solvent: DMAc	• High voltage: 15 kV • Tip-to-collector distance: 20 cm	*Fibers diameter* • D = 600 nm	Oil–water separation Separation efficiency improved from 68% to 88%	[77]
Polymer: nylon-6 Solvent: formic acid	• High voltage: 20–25 kV • Tip-to-collector distance: 20 cm • Dope flowrate: 4.1 μL/min	*Mean fiber diameter* • 250 nm *Nanofiber mass* • 0.007–0.039 g *Total filter mass* • 0.825–0.913 g	Oil–water separation	[78]

(continued)

Table 7.4 (continued)

Polymer–solvent dope	Electrospinning	Fibers characteristics	Separation	Reference
Polymer: styrofoam Solvent: d-Limonene	• High voltage: 20 kV • Tip-to-collector distance: 20 cm	*Fibers diameter* • 300–900 nm *Morphology of fibers* • Beaded fiber structure	N/A	[79]
Polymers: PAN, nylon, polyamide Solvents: DMF, formic acid, DMAc	• High voltage: 20 kV • Needle: hypodermic needle • Collector: flat sheet	*Mean fiber diameters* • PAN: 500 nm • Nylon: 250 nm • Polyamide: 150 nm *Specific area* • PAN: 1.21 m²/g • Nylon: 14.3 m²/g • Polyamide: 22.6 m²/g *Contact angle (water)* • PAN: 38.6° • Nylon: 46.4° • Polyamide: 52.1°	Oil–water separation	[80]

DMAc N,N-Dimethylacetamide, *DMF* N,N-Dimethyl formamide, *PAN* polyacrylonitrile

Like oil–water separation meshes, electrospun membranes can be classified based on the membrane's surface energy, i.e., hydrophobic and oleophilic membranes, or vice versa. Table 7.6 summarizes the recently published works on fabrication of electrospun membranes for oil–water separation. Various polymers have been used to fabricate the microfiltration membranes, specifically for oil–water separation. However, most of these efforts have focused on hydrophilic polymers [92]. This is due to extremely serious fouling tendency for hydrophobic and oleophilic membranes caused by the high viscosity of oil and the small pores in membrane structure. Consequently, for electrospun polymeric membranes the water removing type is especially desired. Hydrophobic and underwater oleophilic nanofibrous membranes are usually fabricated by electrospinning of a hydrophobic polymer, such as PVDF, and extra coating of a secondary layer of a hydrophilic polymeric monomer or chemical group [93]. However, recently, the application of hydrophobic and oleophilic electrospun membranes for oil–water separation is receiving much attention [94]. Moreover, responsive and/or switchable electrospun membranes have also been reported for oily wastewater treatment [95, 96].

Table 7.5 An overview of pressure-driven membrane processes

Process	Membrane material	Membrane geometries	Range of separation	Operating pressure	Reference
Microfiltration	• PTFE • MCE • CA • PP • PE • PES • PS • PVDF	• Flat sheet • Capillary (single or multichannel) • Hollow fiber	• Emulsions • Suspended particles	• 0.5–5 bar	[82]
Ultrafiltration	• CA • PES • PSf[a] • PVDF	• Flat sheet • Capillary (single or multichannel) • Hollow fiber	• Macromolecules	• 3–8 bar	[83]
Nanofiltration	• CA • PES • PSf • PVDF	• Flat sheet • Capillary (single or multichannel) • Hollow fiber • Spiral wound	• Divalent salts	• 10–15 bar	[64]
Reverse osmosis	• CA • PES • PSf	• Flat sheet • Spiral wound	• Monovalent salts	• 15–25 bar	[84]

[a]*PSf* polysulfone

7.5 Discussions and Future Perspectives

7.5.1 *General Discussion*

It is worth noting that for filtration-based oil–water separation, each type of filtering element, i.e., metallic mesh, fabric, coalescing filter, and electrospun membrane, offers a number of unique advantages. Therefore, suitable filtering agent should be selected on the basis of realistic process requirements. In this regard, a number of criteria should be taken into consideration for selective and efficient oil–water separation (see Table 7.7).

Most of the efforts for preparing electrospun filters have been focused on super-hydrophilic filters. This is due to their lower fouling tendency. Moreover, lower operating pressure, mostly close to the atmospheric pressure due to the gravitational driving force, is another advantage of these filters. On the other hand, fabricating nanostructures on a hydrophobic web and/or surface modification to construct superhydrophobic/underwater superoleophilic electrospun filters have been recently investigated. Although many authors have concluded that this strategy is promising for oil–water separation, however, there are still a number of challenges and drawbacks. For instance, superhydrophobic electrospun filters cannot be used under the gravitational driving force [97]. This means adding a pump to the pilot, meaning higher investment. Moreover, due to lower density of oils, fouling tendency is higher in this case.

Table 7.6 Examples of recently published researches on electrospun microfiltration membranes for oil–water separation

Materials	Electrospinning	Features	Performance	Reference
• Nylon 6 • Tetraethyl orthosilicate • Ethyl alcohol (C₂H₅OH) • Polyvinyl acetate (Mw = 140,000) • Ammonium hydroxide (NH₄OH) • Acetone • Sodium dodecyl sulfate • Machine oil (90% base oil with 10% additives, and density of 881.4 kg/m³ at 20 °C) • Formic acid • Acetic acid	• High voltage: 30 kV • Dope flowrate: 0.18 mL/h • Tip-to-collector: 8.8 cm • Relative humidity: 25%	*Analysis* • FE-SEM • TEM • XRD • FTIR • AFM • DLS particle size analyzer • Hydrophobicity *Characteristics* • Contact angle: 15–71° • Porosity: 78% • Pore size: 170 nm • Tensile: 23.3 MPa	• Flux: 4814 LMH/bar • Oil rejection: >98% • Flux recovery: 85%	[65]
• 2,6-Diaminotriptycene • 4,4′-(Hexafluoroisopropylidene) diphthalic anhydride • Pyridine • Acetic anhydride • Methanol • *n*-Hexane • N-methylpyrrolidone	• High voltage: 14 kV • Needle: 22 gauge (0.41 mm ID) • Tip-to-collector: 16 cm • Dope flowrate: 0.5 mL/h • Spinning time: 1 h	*Analysis* • NMR • FTIR • TGA • FE-SEM • Hydrophobicity *Characteristics* • Superhydrophilic • Fibers' diameter: 707 nm • Contact angle: 121°	• Flux: 20–200 L/m² h • Oil rejection: >99%	[87]
• PVDF (Mw = 520,000) • Tetrabutylammonium chloride • Polyethylene polyamine • Acrylic acid • Potassium peroxodisulfate • 1, 2-dichloroethane • Diiodomethane • *n*-Hexane	• High voltage: 30 kV • Needle: pipette • Tip-to-collector: 15 cm • Dope flowrate: 1.0 mL/h • Spinning time: 8 h • Relative humidity: 35%	*Analysis* • Conductivity • FE-SEM • FTIR • BET • XPS • Hydrophobicity • Tensile *Characteristics* • Superhydrophilic • Fibers' diameter: 5–500 nm • Pore size: 0.32 μm • Thickness: 30 μm	• Flux: 9600 L/m² h • Rejection: 99.58%	[88]

(continued)

Table 7.6 (continued)

Materials	Electrospinning	Features	Performance	Reference
• Polyacrylonitrile (Mw = 150,000) • Graphene oxide • DMF • NaOH • HCl	• High voltage: 15 kV • Dope flowrate: 1 mL/h	*Analysis* • AFM • FTIR • XPS • SEM • TEM • Hydrophobicity *Characteristics* • Superhydrophilic	• Flux: 3500 LMH • Oil rejection: 99%	[89]
• Polyacrylonitrile (Mw = 150,000) • DMF • Graphite • Phosphorus pentoxide (P$_2$O$_5$) • Potassium persulfate (K$_2$S$_2$O$_8$) • Sulfuric acid (H$_2$SO$_4$) • Diethylenetriamine (DETA)	• High voltage: 15 kV • Low-voltage: −1.5 kV • Tip-to-collector: 15 cm • Dope flowrate: 1 mL/h • Humidity: 35%	*Analysis* • ATR-FTIR • SEM • AFM • Hydrophobicity *Characteristics* • Superhydrophilic • Fibers' diameter: 0.5–0.9 μm	• Flux: ~10,000 LMH • Oil rejection: ~98%	[90]
• PVDF • Polyamide 6 • DMF • Formic acid • Silicon oil • Tween 80	• High voltage: 10 kV	*Analysis* • SEM • Particle size analyzer • Hydrophobicity *Characteristics* • Pore size: 0.3 μm • Tensile: 7.2 MPa	• Flux: ~900 LMH • Oil rejection: 99%	[91]

7.5.2 Modification of Electrospun Filters

Although electrospun filters have been widely used in various applications, such as water treatment and oily wastewater reclamation, they also have disadvantages such as fouling [98, 99]. As a result, the current trend is to develop new filter materials with special wettable behavior and modified structures, specifically in view of reducing fouling and enhancing flux. One of the promising methods is the modification of common polymers for electrospun filters.

There are three major approaches for modifying the electrospun filters: bulk modification of polymers, surface modification of electrospun filters, and blending, which can also be regarded as surface modification (see Fig. 7.3). Modification methods allow finding a compromise between superhydrophobicity (superoleophilicity) and superhydrophilicity (superoleophobicity). Moreover, it is possible to

Table 7.7 An overview of important criteria for oil–water selective electrospun filters

Requirement	Description	Pros. and Cons.
Superwetting behavior	• Superhydrophilic and superoleophobic filter • Superhydrophobic and superoleophilic filters • Switchable/responsive filters	• Superhydrophilic electrospun filters are the most studied ones, due to their excellent performance • Switchable and responsive electrospun filters have opened a new avenue for oil–water separation
Pore size and its distribution	• Pore size should fit the type of oil–water emulsion • In better words, it should match the oil droplet size • Narrow pore size distribution is crucial for optimum performance	• Metallic meshes and fabrics, due to their large pore size and wide pore size distribution, are not as much efficient as electrospun filters
Antifouling property	• This property is especially desired for the oil-removing filters	
Mechanical strength	• This property is critical for filters' long time performance	• Unfortunately, one of the most important challenges of electrospun filters is their unsuitable mechanical characteristics
Environmental issues	• As an eco-friendly product, recyclability property of a filter is important	• Fortunately, most of electrospun filters for oil–water separation can be fabricated by recyclable materials • Otherwise, they can be used for a long process time, however, by using regular cleaning procedures
Fabrication cost	• Compared to metallic meshes and fabrics, and with investigating the overall performance, electrospun filters are cheaper	

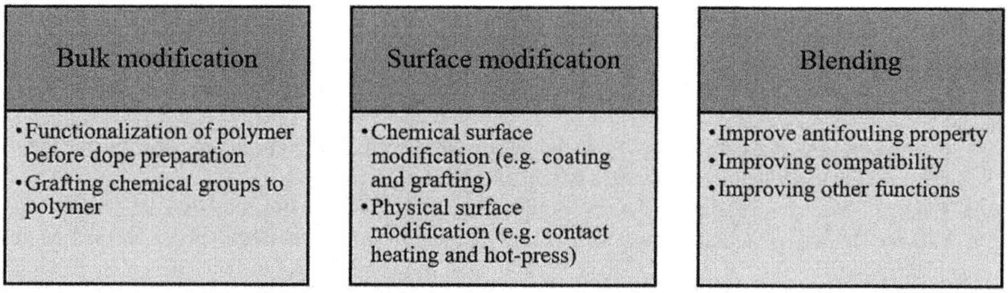

Bulk modification	Surface modification	Blending
• Functionalization of polymer before dope preparation • Grafting chemical groups to polymer	• Chemical surface modification (e.g. coating and grafting) • Physical surface modification (e.g. contact heating and hot-press)	• Improve antifouling property • Improving compatibility • Improving other functions

Fig. 7.3 Three major approaches for modification of electrospun filters

localize the hydrophilic/hydrophobic groups on the fibers' surface, where they have a positive effect on flux enhancement and fouling reduction. Furthermore, the modification can endow functions of filters, such as switchability/responsiveness. Modification of electrospun filters, either in view of increasing superhydrophilicity or superhydrophobicity, can be carried out through a number of techniques. For instance, graft polymerization (attaching hydrophilic/hydrophobic monomers to fibers' surface), plasma treatment (introducing functional groups to the fibers' surface), and physical adsorption of hydrophilic/hydrophobic components to the fibers' surface as well as photo-induced grafting, gamma ray, electron beam-induced grafting are some examples of chemical surface modifications. Further to these chemical procedures, blending, surface coating, and thermal treatment through contact and noncontact heating procedures are examples of physical surface modifications [100–104].

With the development of polymer science and technology, many new polymerization techniques are underdeveloping, such as reversible addition fragmentation chain transfer polymerization [105, 106], as modification of electrospun filters.

7.5.3 Future Strategies

Electrospun filters have shown promising features, not only for oil–water separation but also for other purposes such as water treatment and brine desalination. However, one should note that most of efforts till now have been carried out in bench/lab scale. This is due to a crucial challenge, i.e., the limitation of the electrospinning technique in producing large amount of filters in a short period of time. A promising technique that might develop in future to face this challenge might be gas-assisted electrospinning, also known as electroblowing technique.

References

1. Shirazi MMA, Kargari A (2015) A review on applications of membrane distillation (MD) process for wastewater treatment. J Membr Sci Res 1:101–112
2. Fakhru'l-Razi A, Pendashteh A, Abdullah LC, Awang Biak DR, Madaeni SS, Abidin ZZ (2009) Review of technologies for oil and gas produced water treatment. J Hazard Mater 170:530–551
3. Kargari A, Shirazi MMA (2014) Applications of membrane separation technology for oil and gas produced water treatment. In: Advances in petroleum engineering (vol. 1: Refining). Studium Press LLC, Houston
4. Diya'uddeen BH, Daud WMAW, Aziz ARA (2011) Treatment technologies for petroleum refinery effluents: a review. Process Saf Environ Prot 89:95–105
5. Lin H, Gao W, Meng F, Liao BQ, Leung KT, Zhao L, Chen J, Hong H (2012) Membrane bioreactors for industrial wastewater treatment: a critical review. Critical Rev Environ Sci Technol 42:677–740

6. Gadipelli C, Perez-Gonzalez A, Yadav GD, Ortiz I, Ibanez R, Rathod VK, Marathe KV (2014) Pharmaceutical industry wastewater: review of the technologies for water treatment and reuse. Ind Eng Chem Res 53:11571–11592
7. Hassan AN, Nelson BK (2012) Invited review: anaerobic fermentation of dairy food wastewater. J Dairy Sci 95:6188–6203
8. Beyer J, Trannum HC, Bakke T, Hodson PV, Collier TK (2016) Environmental effects of the Deepwater Horizon oil spill: a review. Mar Pollut Bull 110:28–51
9. Pezeshki SR, Hester MW, Lin Q, Nyman JA (2000) The effects of oil spill and clean-up on dominant US Gulf coast marsh macrophytes: a review. Environ Pollut 108:129–139
10. Spaulding ML (2017) State of the art review and future directions in oil spill modeling. Mar Pollut Bull 115:7–19
11. Yu L, Han M, He F (2013) A review of treating oily wastewater. Arab J Chem 10:S1913–S1922. https://doi.org/10.1016/j.arabjc.2013.07.020
12. Zhou YB, Tang XY, Hu XM, Fritschi S, Lu J (2008) Emulsified oily wastewater treatment using a hybrid-modified resin and activated carbon system. Sep Purif Technol 63:400–406
13. Jamaly S, Giwa A, Hasan SW (2015) Recent improvements in oily wastewater treatment: progress, challenges, and future opportunities. J Environ Sci 37:15–30
14. Emamjomeh MM, Sivakumar M (2009) Review of pollutants removed by electrocoagulation and electrocoagulation/floatation processes. J Environ Manag 90:1663–1679
15. Golestanbagh M, Parvini M, Pendashteh A (2016) Integrated systems for oilfield produced water treatment: the state of the art. Energy Sources A 38:3404–3411
16. Moussa DT, El-Naas MH, Nasser M, Al-Marri MJ (2017) A comprehensive review of electrocoagulation for water treatment: potentials and challenges. J Environ Manag 186:24–41
17. Padaki M, Murali RS, Abdullah MS, Misdam N, Moslehyani A, Kassim MA, Hilal N, Ismail AF (2015) Membrane technology enhancement in oil-water separation. A review. Desalination 357:197–207
18. Pintor AMA, Vilar VJP, Botelho CMS, Boaventura RAR (2016) Oil and grease removal from wastewaters: sorption treatment as an alternative to state-of-the-art technologies. A critical review. Chem Eng J 297:229–255
19. Munirasu S, Haija MA, Banat F (2016) Use of membrane technology for oil field and refinery produced water treatment-a review. Process Saf Environ Prot 100:183–202
20. Xue Z, Cao Y, Liu N, Feng L, Jiang L (2014) Special wettable materials for oil/water separation. J Mater Chem A 2:2445–2460
21. Shirazi MMA, Kargari A, Tabatabaei M (2014a) Evaluation of commercial PTFE membranes in desalination by direct contact membrane distillation. Chem Eng Process 76:16–25
22. Ray SS, Chen SS, Li CW, Nguyen NC, Nguyen HT (2016) A comprehensive review: electrospinning technique for fabrication and surface modification of membranes for water treatment application. RSC Adv 6:85495–85514
23. Lalia BS, Kochkodan V, Hashaikeh R, Hilal N (2013) A review on membrane fabrication: structure, properties and performance relationship. Desalination 326:77–95
24. Ahmed F, Lalia BS, Kochkodan V, Hilal N, Hashaikeh R (2016) Electrically conductive polymeric membranes for fouling prevention and detection: a review. Desalination 391:1–15
25. Bet-moushoul E, Mansourpanah Y, Farhadi K, Tabatabaei M (2016) TiO$_2$ nanocomposite based polymeric membranes: a review on performance improvement for various applications in chemical engineering processes. Chem Eng J 283:29–46
26. Otitoju TA, Amad AL, Ooi BS (2016) Polyvinylidene fluoride (PVDF) membrane for oil rejection from oily wastewater: a performance review. J Water Process Eng 14:41–59
27. Paul M, Jons SD (2016) Chemistry and fabrication of polymeric nanofiltration membranes: a review. Polymer 103:417–456
28. Zheng X, Zhang Z, Yu D, Chen X, Cheng R, Min S, Wang J, Xiao Q, Wang J (2015) Overview of membrane technology applications for industrial wastewater treatment in China to increase water supply. Resour Conserv Recycl 105:1–10

29. Ma Q, Cheng H, Fane AG, Wang R, Zhang H (2016) Recent development of advanced materials with special wettability for selective oil/water separation. Small 12:2186–2202
30. Onda T, Shibuichi S, Satoh N, Tsujii K (1996) Super-water-repellent fractal surfaces. Langmuir 12:2125–2127
31. Öner D, McCarthy TJ (2000) Ultrahydrophobic surfaces. Effects of topography length scales on wettability. Langmuir 16:7777–7782
32. Ma M, Hill RM (2006) Superhydrophobic surfaces. Curr Opin Colloid Interface Sci 11:193–202
33. Shirazi MMA, Kargari A, Bazgir S, Tabatabaei M, Shirazi MJA, Abdullah MS, Matsuura T, Ismail AF (2013) Characterization of electrospun polystyrene membrane for treatment of biodiesel's water-washing effluent using atomic force microscopy. Desalination 329:1–8
34. Wu J, Wang N, Wang L, Dong H, Zhao Y, Jiang L (2012) Electrospun porous structure fibrous film with high oil adsorption capacity. Appl Mater Interfaces 4:3207–3212
35. Sarbatly R, Krishnaiah D, Kamin Z (2016) A review of polymer nanofibers by electrospinning and their application in oil-water separation for cleaning up marine oil spills. Mar Pollut Bull 106:8–18
36. Rohrbach K, Li Y, Zhu H, Liu Z, Dai J, Andreasen J, Hu L (2014) A cellulose based hydrophilic, oleophobic hydrated filter for water/oil separation. Chem Commun 50:13296–13299
37. Yang J, Zhang Z, Xu X, Zhu X, Men X, Zhou X (2012) Superhydrophilic-superhydrophobic coatings. J Mater Chem 22:2834–2837
38. Liu K, Jiang L (2011) Metallic surfaces with special wettability. Nanoscale 3:825–838
39. Feng L, Zhang Z, Mai Z, Ma Y, Liu B, Jiang L, Zhu D (2004) A super-hydrophobic and super-oleophilic coating mesh film for the separation of oil and water. Angew Chem Int Ed 43:2012–2014
40. Wang F, Lei S, Li C, Ou J, Xue M, Li W (2014) Superhydrophobic Cu mesh combined with a superoleophilic polyurethane sponge for oil spill adsorption and collection. Ind Eng Chem Res 53:7141–7148
41. Yang J, Tang Y, Xu J, Chen B, Tang H, Li C (2015) Durable superhydrophobic/superoleophilic epoxy/attapulgite nanocomposite coatings for oil/water separation. Surf Coat Technol 272:285–290
42. Lee CH, Johnson N, Drelich J, Yap YK (2011) The performance of superhydrophobic and superoleophilic carbon nanotube meshes in water-oil filtration. Carbon 49:669–676
43. Zhang F, Zhang WB, Shi Z, Wang D, Jin J, Jiang L (2013) Nanowire-haired inorganic membranes with superhydrophilicity and underwater ultralow adhesive superoleophobicity for high-efficiency oil/water separation. Adv Mater 25:4192–4198
44. Liu N, Chen Y, Lu F, Cao Y, Xue Z, Li K, Feng L, Wei Y (2013) Straightforward oxidation of a copper substrate produces an underwater superoleophobic mesh for oil/water separation. ChemPhysChem 14:3489–3494
45. Wang B, Guo Z (2013) pH-responsive bidirectional oil-water separation material. Chem Commun 49:9416–9418
46. Wang B, Li J, Wang G, Liang W, Zhang Y, Shi L, Guo Z, Liu W (2013) Methodology for robust superhydrophobic fabrics and sponges from in situ growth of transition metal/metal oxide nanocrystals with thiol modification and their applications in oil/water separation. ACS Appl Mater Interfaces 5:1823–1839
47. Zhou X, Zhang Z, Xu X, Guo F, Zhu X, Men X, Ge B (2013) Robust and durable superhydrophobic cotton fabrics for oil/water separation. ACS Appl Mater Interfaces 5:7208–7214
48. Gupta RK, Dunderdale GJ, England MW, Hozumi A (2017) Oil/water separation techniques: a review of recent progresses and future directions. J Mater Chem A 5:16025–16058. https://doi.org/10.1039/C7TA02070H
49. Ramakrishna S, Shirazi MMA (2015) Electrospun membranes: next generation membranes for desalination and water/wastewater treatment. J Membr Sci Res 1:46–47
50. Shirazi MMA, Kargari A, Ramakrishna S, Doyle J, Rajendarian M, Babu PR (2017) Electrospun membranes for desalination and water/wastewater treatment: a comprehensive review. J Membr Sci 3:209–227

51. Kulkarni A, Bambole VA, Mahanwar PA (2010) Electrospinning of polymers, their modeling and applications. Polym Plast Technol Eng 49:427–441
52. Sas I, Gorga R, Joines J, Thoney KA (2012) Literature review on superhydrophobic self-cleaning surfaces produced by electrospinning. J Polymer Sci B 50:824–845
53. Tijing LD, Choi JS, Lee S, Kim SH, Shon HK (2014) Recent progress of membrane distillation using electrospun nanofibrous membrane. J Membr Sci 453:435–462
54. Konwarch R, Karak N, Misra M (2013) Electrospun cellulose acetate nanofibers: the present status and gamut of biotechnological applications. Biotechnol Adv 31:421–437
55. McKee MG, Wilkes GL, Colby RH, Long TE (2004) Correlations of solution rheology with electrospun fiber formation of linear and branched polyesters. Macromolecules 37:1760–1767
56. Frenot A, Chronakis IS (2003) Polymer nanofibers assembled by electrospinning. Curr Opin Colloid Interface Sci 8:64–75
57. Kang G, Cao Y (2014) Application and modification of poly(vinylidene fluoride) (PVDF) membranes-a review. J Membr Sci 463:145–165
58. Su CI, Shih JH, Huang MS, Wang CM, Shih WC, Liu YS (2012) A study of hydrophobic electrospun membrane applied in seawater desalination by membrane distillation. Fibers Polym 13:698–702
59. Chen B, Qiu J, Sakai E, Kanazawa N, Liang R, Feng H (2016) Robust and superhydrophobic surface modification by a "paint + adhesive" method: applications in self-cleaning after oil contamination and oil-water separation. ACS Appl Mater Interfaces 8:17659–17667
60. Ahmed FE, Lalia BS, Hashaikeh R (2017) Membrane-based detection of wetting phenomenon in direct contact membrane distillation. J Membr Sci 535:89–93
61. Mirtalebi E, Shirazi MMA, Kargari A, Tabatabaei M, Ramakrishna S (2014) Assessment of atomic force and scanning electron microscopes for characterization of commercial and electrospun nylon membranes for coke removal from wastewater. Desalin Water Treat 52:6611–6619
62. Ahmed FE, Lalia BS, Hashaikeh R (2015) A review on electrospinning for membrane fabrication: challenges and applications. Desalination 356:15–30
63. Eykens L, De Sitter K, Dotremond C, Pinoy L, Van der Bruggen B (2017) Membrane synthesis for membrane distillation: a review. Sep Purif Technol 182:36–51
64. Mohammad AW, Teow YH, Ang WL, Chung YT, Oatley-Radcliffe DL, Hilal N (2015) Nanofiltration membranes review: recent advances and future prospects. Desalination 356:226–254
65. Islam MS, McCutcheon JR, Rahaman MS (2017) A high flux polyvinyl acetate-coated electrospun nylon 6/SiO$_2$ composite microfiltration membrane for the separation of oil-in-water emulsion with improved antifouling performance. J Membr Sci 537:297–309
66. Ashrafizadeh SN, Motaee E, Hoshyargar V (2012) Emulsification of heavy crude oil in water by natural surfactants. J Pet Sci Eng 86:137–143
67. Eow JS, Ghadiri M (2002) Electrostatic enhancement of coalescence of water droplets in oil: a review of the technology. Chem Eng J 85:357–368
68. Hakansson A (2016) Experimental methods for measuring coalescence during emulsification-a critical review. J Food Eng 178:47–59
69. ACS Report, Liquid-liquid coalesce design manual, ACS Industries LP, USA (Report code: 800-231-007) (http://people.clarkson.edu/~wwilcox/Design/coalesc.pdf)
70. Li J, Gu Y (2005) Coalescence of oil-in-water emulsions in fibrous and granular beds. Sep Purif Technol 42:1–13
71. Sokolovic RMS, Govedarica DD, Sokolovic DS (2010) Separation of oil-in-water emulsion using two coalescers of different geometry. J Hazard Mater 175:1001–1006
72. Sokolovic RMS, Vulic TJ, Sokolovic SM (2007) Effect of length on steady-state coalescence of oil-in-water emulsion. Sep Purif Technol 56:79–84
73. Hong A, Fane AG, Burford R (2003) Factors affecting membrane coalescence of stable oil-in-water emulsions. J Membr Sci 222:19–39

74. Shirazi MJA, Bazgir S, Shirazi MMA, Ramakrishna S (2013b) Coalescing filtration of oily wastewaters: characterization and application of thermal treated, electrospun polystyrene filters. Desalin Water Treat 51:5974–5986
75. Shin C, Chase GG (2006) Separation of water-in-oil emulsions using glass fiber media augmented with polymer nanofibers. J Dispers Sci Technol 27:517–522
76. Shin C (2006) Filtration application from recycled expanded polystyrene. J Colloid Interface Sci 302:267–271
77. Shin C, Chase GG, Reneker DH (2005a) Recycled expanded polystyrene nanofibers applied in filter media. Colloids Surf A Physicochem Eng Asp 262:211–215
78. Shin C, Chase GG, Reneker DH (2005b) The effect of nanofibers on liquid-liquid coalescence filter performance. AICHE J 51:3109–3113
79. Shin C, Chase GG (2005) Nanofibers from recycle waste expanded polystyrene using natural solvent. Polym Bull 55:209–215
80. Shin C, Chase GG (2004) Water-in-oil coalescence in micro-nanofiber composite filters. AICHE J 50:343–350
81. Shirazi MJA, Bazgir S, Shirazi MMA (2014b) Edible oil mill effluent; a low-cost source for economizing biodiesel production: electrospun nanofibrous coalescing filtration approach. Biofuel Res J 1:39–42
82. Rad SN, Shirazi MMA, Kargari A, Marzban R (2016) Application of membrane separation technology in downstream processing of Bacillus thuringiensis biopesticide: a review. J Membr Sci Res 2:66–77
83. Shi X, Tal G, Hankins NP, Gitis V (2014) Fouling and cleaning of ultrafiltration membranes: a review. J Water Process Eng 1:121–138
84. Khulbe KC, Matsuura T (2017) Recent progresses in preparation and characterization of RO membranes. J Membr Sci Res 3:174–186. https://doi.org/10.22079/jmsr.2016.22147
85. Barhate RS, Ramakrishna S (2007) Nanofibrous filtering media: filtration problems and solutions from tiny materials. J Membr Sci 296:1–8
86. Sundarrajan S, Ramakrishna S (2013) New directions in nanofiltration applications-are nanofibers the right materials as membranes in desalination? Desalination 308:198–208
87. Zhai TL, Du Q, Xu S, Wang Y, Zhang C (2017) Electrospun nanofibrous membrane of porous fluorine-containing triptycene-based polyimides for oil/water separation. RCS Adv 7:22548–22552
88. Cheng B, Li Z, Li Q, Ju J, Kang W, Naebe M (2017) Development of smart poly(vinylidene fluoride)-graft-poly(acrylic acid) tree-like nanofiber membrane for pH-responsive oil/water separation. J Membr Sci 534:1–8
89. Zhang J, Pan X, Xue Q, He D, Zhu L, Guo Q (2017a) Antifouling hydrolyzed polyacrylonitrile/graphene oxide membrane with spindle-knotted structure for highly effective separation of oil-water emulsion. J Membr Sci 532:38–46
90. Zhang J, Xue Q, Pan X, Jin Y, Lu W, Ding D (2017b) Graphene oxide/polyacrylonitrile fiber hierarchical-structured membrane for ultra-fast microfiltration of oil-water emulsion. Chem Eng J 307:643–649
91. Lv R, Yin M, Zheng W, Na B, Wang B (2017) Poly(vinylidene fluoride) fibrous membranes doped with polyamide 6 for highly efficient separation of a stable oil/water emulsion. J Appl Polym Sci 134:44980–44985
92. Obaid M, Barakat NAMAA, Khalil KA (2015) Effective and reusable oil/water separation membranes based on modified polysulfone electrospun nanofiber mats. Chem Eng J 259:449–456
93. Yuan T, Meng J, Hao T, Wang Z, Zhang Y (2015) A scalable method toward superhydrophilic and underwater superoleophobic PVDF membranes for effective oil/water emulsion separation. ACS Appl Mater Interfaces 7:14896–14904
94. Zhou Z, Wu XF (2015) Electrospinning superhydrophobic-superoleophilic fibrous PVDF membranes for high-efficiency water-oil separation. Mater Lett 160:423–427

95. Che H, Huo M, Peng L, Fang T, Liu N, Feng L, Wei Y, Yuan J (2015) CO_2-responsive nanofibrous membranes with switchable oil/water wettability. Angew Chem 127:9062–9066
96. Li JJ, Zhu LT, Luo ZH (2016) Electrospun fibrous membrane with enhanced switchable oil/water wettability for oily water separation. Chem Eng J 287:474–481
97. Viswanadam G, Chase GG (2013) Water-diesel secondary dispersion separation using superhydrophobic tubes of nanofibers. Sep Purif Technol 104:81–88
98. Gao W, Liang H, Ma J, Han M, Chen ZL, Han ZS, Li GB (2011) Membrane fouling control in ultrafiltration technology for drinking water production: a review. Desalination 272:1–8
99. Guo W, Ngo HH, Li J (2012) A mini-review on membrane fouling. Bioresour Technol 122:27–34
100. Carlmark A, Larsson E, Malmstrom E (2012) Grafting of cellulose by ring-opening polymerization – a review. Eur Polym J 48:1646–1659
101. Hokkanen S, Bhatnagar A, Sillanpaa M (2016) A review on modification methods to cellulose-based adsorbents to improve adsorption capacity. Water Res 91:156–173
102. Kang H, Liu R, Huang Y (2015) Graft modification of cellulose: methods, properties and applications. Polymer 70:A1–A16
103. Kochkodan V, Hilal N (2015) A comprehensive review on surface modified polymer membranes for biofouling mitigation. Desalination 356:187–207
104. Zhao C, Xue J, Ran F, Sun S (2013) Modification of polyethersulfone membranes – a review of methods. Prog Mater Sci 58:76–150
105. Khulbe KC, Feng C, Matsuura T (2010) The art of surface modification of synthetic polymeric membranes. J Appl Polym Sci 115:855–895
106. Rana D, Matsuura T (2010) Surface modifications for antifouling membranes. Chem Rev 110:2448–2471

Chapter 8
Affinity Membranes for Capture of Cells and Biological Substances

Rameshkumar Saranya, Rajendiran Murugan, Manasa Hegde, James Doyle, and Ramesh Babu

Abstract Fundamental approaches of developing electrospun affinity membranes and their applications in cell capture and separation of biologically active substances are discussed. Basic principles of affinity membrane-based separation process and types of various surface modifications are described, to produce electrospun affinity nanofibers for filtration or fractionation of desired biological substances. The relevant advancements in electrospinning for novel surface modifications have been discussed with a special focus on coupled ligands and molecular imprinted polymers. Applications pertaining to cell capturing and filtration of biomolecules are reviewed in detail, with reference to recent reports on electrospun affinity nanofiber membranes for various antibody and protein purification.

R. Saranya · M. Hegde
AMBER, CRANN Institute, Trinity College Dublin, Dublin, Republic of Ireland

School of Physics, Trinity College Dublin, Dublin, Republic of Ireland

R. Murugan
OEM Diagnostics, Merck Ltd, County Cork, Republic of Ireland

AMBER, CRANN Institute, Trinity College Dublin, Dublin, Republic of Ireland

J. Doyle
AMBER, CRANN Institute, Trinity College Dublin, Dublin, Republic of Ireland

R. Babu (✉)
AMBER, CRANN Institute, Trinity College Dublin, Dublin, Republic of Ireland

School of Physics, Trinity College Dublin, Dublin, Republic of Ireland

BEACON - Bioeconomy Research Centre, University College Dublin, Dublin, Republic of Ireland
e-mail: babup@tcd.ie

© Springer International Publishing AG, part of Springer Nature 2018
M. L. Focarete et al. (eds.), *Filtering Media by Electrospinning*,
https://doi.org/10.1007/978-3-319-78163-1_8

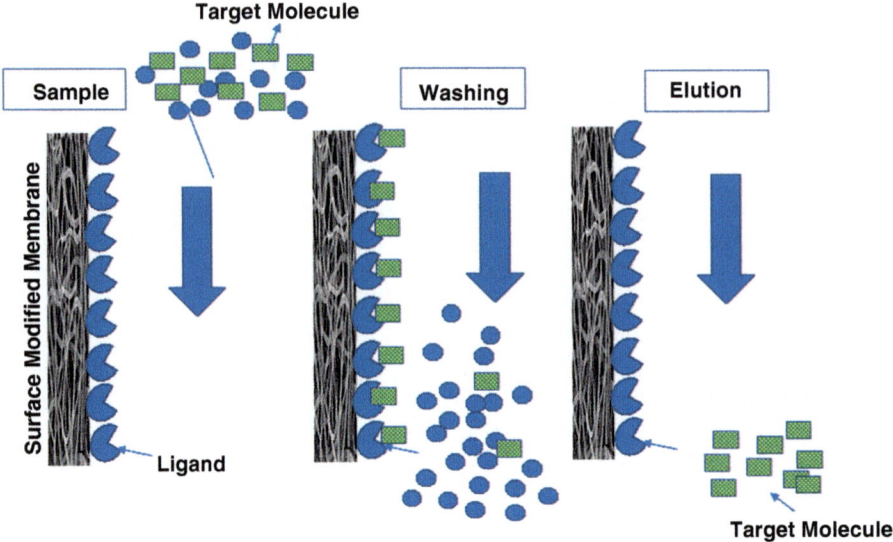

Fig. 8.1 Schematic representation of affinity membrane separation process

8.1   Introduction

Affinity membranes are a broad class of membranes that selectively capture targeted molecules for use in such technological advances as fixed-bed liquid chromatography and membrane filtration. This is primarily achieved by immobilizing a specific molecule or ligand onto the membrane surface [1, 2] thereby permitting the purification of molecules based on differences in physical/chemical properties or biological functions, rather than purely molecular weight or size. Common ligands used in affinity separation membranes include dye ligands, antigen–antibody ligands, ion-exchange ligands, and other biological ligands such as protein libraries and enzymes, depending on the requirement. Combining both the high productivity associated with membranes and the outstanding selectivity of the chromatography resins, affinity membrane chromatography is now an attractive and competitive method for purifying proteins, or other biomolecules, from biological fluids [3–8]. Figure 8.1 represents the scheme of affinity separation process for the conventional membranes.

Affinity membranes were developed primarily for purifying proteins and other biomolecule forms from complex biologically fluids. These membranes combine both the outstanding selectivity of the chromatography resins and overcome the reduced pressure drops associated with filtration membranes [9, 10].

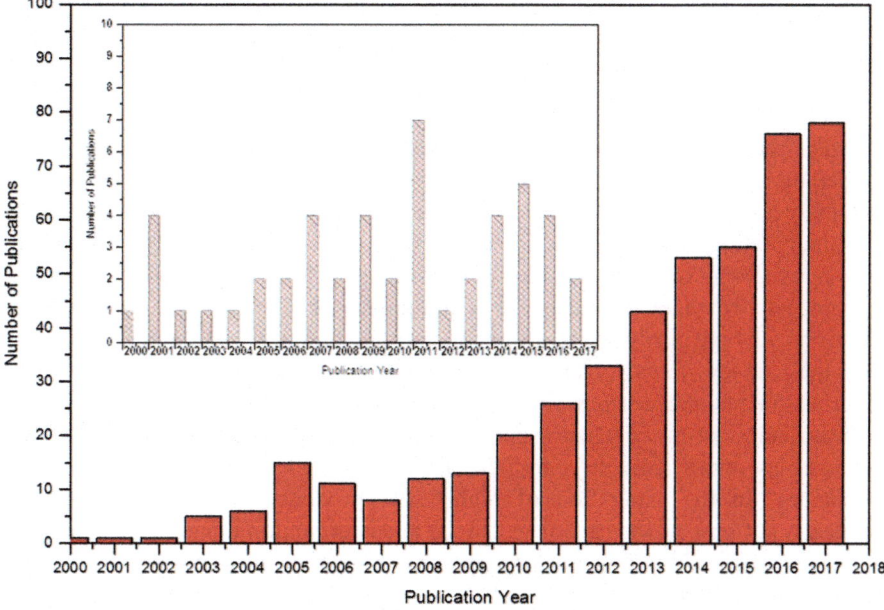

Fig. 8.2 Graph presenting yearly numbers of research reports on electrospun membranes (2000–2017) for filtration. *Inset:* Number of publications on electrospun affinity membranes for biological filtration

8.2   Overview of Affinity Membrane Separation

Affinity membrane-based separation techniques and use of electrospun membranes are becoming increasingly popular in biotechnology, food, and pharmaceutical industries due to their versatility and hydrodynamic benefits over traditional materials and methods. Many applications are emerging for electrospun nanofiber membranes in the field of filtration and recovery of biological substances. Since 2010, there has been a rapid increase in the number of publications investigating electrospun membranes for filtration applications, as shown in Fig. 8.2 [11]. Similar trends were observed for electrospun affinity membranes for cell capture and biological substances, as shown in the inset of Fig. 8.2. The majority of technological activity in the affinity membrane area takes place in China, Southeast Asia, Australia, New Zealand, and the USA and taken together approximately 85% of publications come from these regions.

The operation mode of membranes and affinity-based sorption, which basically works for selective sieving characteristics and site-specific adsorption, are similar to the standard chromatographic procedures: adsorption, washing, and elution but with an additional cleaning step. Washing is generally carried out via *back flushing* through the membrane, whereas elution depends on the mode of operation:

1. Back flushing is utilized to capture particle-free suspensions
2. Forward filtering is employed for particle-rich suspensions.

The isolation and recovery of many biologically active materials in low concentration are often complex. Since the functional properties of biological materials critically depend on their structure, the permissible changes with temperature, pH, ionic strength, and solvent used for the purification process are extremely restricted. Thus even basic techniques such as crystallization, evaporation, and precipitation cannot be employed, because they may ultimately decrease the biological activity of the products, due to physical and chemical interactions during the process. To meet the stringent requirement for the recovery of biological products with high purity levels, affinity membrane-based technologies are evolving. Over the past few years, advancements in membrane design and in new polymeric materials for membranes as well as the evolution of new generation of membranes such as nanofiber-based electrospun membranes, have made affinity membrane technologies more attractive for the isolation and purification of biological substances. In the separation of high value products, membrane separations frequently suffer from the lack of specificity and hence chromatography is employed as an alternative. Utilizing the principles of biological interactions, the method referred to as "affinity chromatography" was developed, commonly based on packed bed sorbents. Yet there are limitations associated with pack beds, notably low flow rates and diffusional mass transfer, which can be overcome by the use of polymeric membranes. The necessity of high specificity and improved flow rate and mass transfer for high-resolution separations has driven the need to combine the principle of chromatography with membrane filtration, yielding the concept of affinity membrane technology.

8.3   Electrospinning: Versatile Technology for Affinity Membrane Preparation

Compared to traditional membrane production techniques, electrospinning (ES) is considered to be one of the most versatile methods emerging in recent years for the production of thin polymeric fibrous membranes [12], for multiple applications, including the capture and recovery of cells and bioactive compounds [13]. ES is a well-known technique by which an external electric field is applied to a polymer solution or polymer melt, leading to the formation of polymer fibers. The ES method allows for high volume production of lightweight, highly functional mesh-like structures. This versatile *electrohydrodynamic* technique results in products with a controlled porous structure, which can be prepared in film form, as a coating or with a multidimensional network structure (from micron to nanoscale size). In its most basic form, the ES process involves placing a polymer solution in a pipette between two electrodes, which can create a large potential difference (in the kV regime). This large voltage electrostatically draws the polymer solution towards a grounded target in a thin, continuous jet, leading to a deposition of a fibrous web. Starting from a specific polymeric solution, the resulting pore size and pore distribution of the obtained web can be controlled by varying instrumental parameters such as the distance between the spinneret and the collector, voltage, collector type (static or rotated dynamically), and jet spindle speed. In addition to the optimization of system parameters, advances in higher throughput ES system design continue to be developed [14–19].

ES membranes are characterized by very large surface area-to-volume ratio, high porosity with nano-/microscale pores, and continuous interconnectivity, and therefore offer huge potential for incorporating various ligands to create affinity membranes. A multitude of electrospun nanofiber designs enables the easy functionalization of nanofiber membrane surfaces [20, 21].

8.4   Benefits of Electrospun Affinity Membranes over Traditional Methods

The advent of nanotechnology is leading to breakthrough advancements in a variety of applications, in particular, in the biomedical and pharmaceutical fields. Inspired by nanotechnology, affinity membrane structures have recently been developed, in the form of electrospun nanofibers offering huge commercial potential in the biological sector. Moreover, fabrication of fibers incorporating functional nanomaterials/nanostructures is also possible by adopting ES as the fabrication technique [22, 23]). The benefits of ES include, synthesizing fibers of controlled morphology, enhanced surface area, encapsulation efficiency, usability, and reproducibility of the final product on a large scale [24]. The use of ES for nanotechnology-based medical devices and sensors is a preferred one and can be considered commercially competitive when compared to existing methods of sensor-based technology and therapeutic medical systems.

Most of the research activities aimed at biological scaffolds and media/cell filtration are being explored with the help of extensive applications of ES technology [25–27]. The promising features of ES designed nanofibers could revolutionize the field of drug delivery and other significant medical applications, thus enabling desirable modifications of products in physical property, synergistic delivery mechanism, compatibility, and so on [28]. Electrospun nanofibrous membranes are preferred over other conventional membranes because they simultaneously achieve adequate high strength and biomechanical stability to resist wearing and tearing at edges of developed nanofibrous affinity membranes. As affinity separation processes involve filtration media which would be exposed to extreme temperature, mechanical shock, vibration, shear/stress, and reactions during the fluid flow [29], acceptable mechanical, chemical, and physical stabilities all play a huge role in determining the filtering efficiency. The ES method is considered an effective approach to fulfill the afore mentioned integrity and stability criteria of affinity nanofibrous membranes. However, there are several processing conditions with respect to polymer, process/operation, and ambient conditions which influence the ability of ES to extrude nanofibers of necessary functionality, stability, and efficiency. The future commercialisation of affinity membranes will require continued materials improvements, and research is focusing on novel geometries of electrospun nanofibers (such as hollow nanofibers, nanofiber mats/sheets) to provide the next generation systems for areas such as gene therapy, chemotherapy, and immunotherapy.

8.5 Electrospinning for Affinity Membrane Preparation

By judiciously altering the electrospun parameters, it is possible to fabricate fibers with different morphology from various starting systems, such as metals, ceramics, and polymers. Also, the modification of starting materials is possible by incorporation of novel nanomaterials, bioactive molecules, ligands, and even living cells for specific diagnosis and point of care therapies [30]. Affinity membrane preparation based on electrospun nanofibers is advantageous because ES tends to achieve higher ratio of surface area to volume, which is the most desired feature of an ideal affinity membrane.

With regard to different steps involved in affinity membrane preparation, the following sections discuss the base nanofibrous membrane development by ES. Other activation and immobilization methods are quite comparable to conventional methods of affinity membrane preparation.

8.5.1 ES of Polymeric Nanofibers

The assembly of polymeric nanofiber mesh via ES methods typically exhibits micro-/nanoscaled pore diameter with high porosity and interconnectivity of the interstitial space, thus eventually resulting in greater surface area than conventional filter media or fibers. Evidence of research focusing on developing electrospun nanofibers for affinity-based separation can be found in the literature. In addition, studies related to the use of electrospun membranes for self-assembling of desired ligands/molecules on the polymer matrix exist, highlight the remarkable multifunctionality and suitability for biological media filtration [27, 31]. Biomimetic membranes carrying therapeutic drug for point-of-delivery and providing stimuli-responsive ability to biological nanofibers are also another domain of research interest for electrospun nanofibers [32–34]. However, such topics are beyond the scope of the present chapter.

Generally, bulk polymers including polyethylene terephthalate (PET) [35], polysulfone (PSF) [3, 4], cellulose [36], cellulose acetate [37], polyacrylonitrile (PAN) [38, 39], and polyamide [40, 41] are used to develop electrospun nanofibrous membrane materials for affinity binding. However, polymer blends are mostly preferred in order to achieve a synergistic effect and to efficiently tailor both pore structure and binding sites [42, 43]. In the case of blending one or more polymers, the bulk polymer matrix is responsible for the pore (membrane) morphology, whereas non-covalent interaction is exerted by means of a "template"-like functional polymer. Hence, the preparation of affinity membranes by ES follows two broad approaches based on: a) introducing additional functional layers and b) introducing responsiveness (either self or induced). These approaches help not only in surface functionality but also in preserving the bulk polymer structure (base membrane nanofibers) owing to non-covalent interactions with the bulk polymers [44].

8.5.2 Activation and Immobilization

Activation of the electrospun fibers helps in facilitating the binding/coupling of the desired functional groups and molecule-specific ligands, aimed at specific applications requiring low-resolution selectivity and high-resolution selectivity, respectively. Activation is generally referred as "modification" of the surface of the nanofibers in the present context, introducing affinity-based interactions which enable the desired separation and prevent unwanted interactions (secondary adsorption/adhesion) [44]. Controlled reaction/interaction of the permeating (bio)molecules/cells acts as a key for affinity separation. Most of the affinity membranes operate on surface-selective separation for which activation is considered as an important step. In order to link the site-specific affinity ligands either directly or indirectly to the membrane matrix, an activation step is essential to retain the general membrane properties (mechanical resistance and stability). Several modification reactions to activate the polymeric membrane containing various functional groups have been reviewed [45, 46]. Clearly, the activation reactions are specific to each functional groups, for example, carbamylation and thiocarbamylation reactions for polymers containing amine-functional groups [47]. The influence of the activation step must not hamper the specific binding affinity of the ligands, and could in fact facilitate surface modification to ensure the attachment of ligand to the support polymeric nanofiber.

Practically, the method of activation follows forced convection in the case of hydrophilic polymeric substrate, wherein the activation reactions have employed nonpolar solvents to improve the membrane wettability to induce biological compatibility of the polymeric nanofiber. The accessibility of the ligand to the substrate can also be enhanced by means of placing *spacer molecules*, especially when direct coupling of ligand deems ineffective. Examples of spacer molecules include diamines and dihydrazines, which offer molecular interactions between the immobilized ligand and the desired target molecules. These spacer molecules are mostly bifunctional, suitable to interact both with the polymeric substrate and the ligand molecules [48].

Immobilization is one of the important preparative steps, which often refers to surface modification or functionalization, which can range from passive adsorption to sophisticated covalent binding or cross-linking methods using chemically activated polymers. Immobilization is the method of reversibly binding of the

Table 8.1 Classification of immobilization ligands on electrospun affinity membranes

Ligand type	Examples	Method/Source	Advantages	Disadvantages
Synthetic	• Molecule-specific dyes • Nonnatural peptides	• De novo synthesis • Rational selection from ligand libraries • Modification of molecular structures	• High capacity • Low cost • High stability	• Toxicity • Nominal selectivity
Biological	• Antigen–antibodies • RNA and DNA fragments (aptamers) • Binding or receptor proteins	• Naturally derived • In vitro from biological systems	• Low toxicity • Very-high selectivity	• High cost • Low stability

ligands that are specific to a molecule, or group of molecules, to carry out purification/separation by means of affinity reactions [49]. The selection of ligands plays a major role in determining the specificity and stability of the affinity membrane and separation performance [50]. There are different classes of ligands including dyes, amino acids, peptides, proteins (binding or receptor), enzymes, antibodies and antigens, coenzyme, RNA and DNA strands, vitamins, nucleotides, metal chelates, and so on [51]. Broadly, ligands for affinity-based membrane separation are classified into biological and synthetic. Differences between synthetic and biological ligands in terms of method of derivation along with their merits and demerits are summarized in Table 8.1.

Although ligand selection is based on the application and on polymeric substrate, there are many factors to consider such as toxicity, stability, cost, and so on. Ligand selection must be based on its capacity to selectively and reversibly bind to the desired biological substance to be isolated. Moreover, ligands having additional functional groups are of significant interest as this enables activation procedures to get attached to the support material, say polymeric nanofibers. Different subclasses of synthetic and biological ligands are also described below.

8.5.2.1   Immobilization of Low-Resolution Functional Groups

Immobilization of relevant functional groups onto nanofiber surfaces is especially attractive if the resulting membrane could be used for affinity binding. Immobilization of functional groups such as carboxyl, amine, alkyl, hydroxyl, amides, ketones, and urethanes is typically based on simple surface chemistry and the inherent chemical affinity towards specific molecules. For example, most of the polysaccharides, such as alginate, chitosan, agarose, and cellulose, are found to have inherent ionic polymerization ability and can be modified based on methods employing cyanogen bromide, triazine, and carbonylation [46]. The ionic polymerization effect (coacervation) of alginate and chitosan has been mainly utilized for immobilization and purification of variety of enzymes [52]. Moreover, some of the synthetic polymers containing carboxyl or sulfone groups are known for nonpolar adsorption. Electrospun nanofibers from polystyrene were found to display selectivity for basic and neutral solutes, whereas sulfonated polyetheretherketone (SPEEK) possesses inherent separation capacity for cationic solutes owing to its specificity towards monovalent cations [53]. Apart from surface functional groups, coordination chemistry is yet another approach for developing affinity membranes. This is because, in affinity binding, the ligands, even when placed at a small distance away from the surface of the substrate nanofiber, could result in binding due to specific interactions. For instance, electrospun porphyrin copolymer nanofibers were naturally found to have inherent covalent interactions [54]. Such nanofibers could aid the recovery of desired phenolics and aromatics by means of π–π interactions. Applications pertaining to metal-ion affinity-based separation make use of such covalent interactions for selective separation of proteins. Moreover, electrospun fibers with rare earth-based emitters immobilized can also be applied for detection of gaseous molecules; however, it is beyond the scope of affinity-based cell filtration and biomolecule separation which are mostly dealt in this chapter.

8.5.2.2 Immobilization of High-Resolution Ligands

Synthetic and Pseudobiospecific Ligands

High-resolution affinity-based separations are performed by means of two broad classification of ligands as mentioned in Table 8.1. Most of the affinity-based sensing utilizes synthetic ligands due to the benefits of high stability and ease of immobilization, at a lower cost. Limitations related to the use of synthetic ligands however exist, such as their toxicity and poor selectivity for immunoaffinity and other diagnostic applications. Biological ligands, having the ability to bind very specifically to appropriate biomolecules, still have drawbacks related to large-scale application and complexity in immobilization. Synthetic ligands classified under "pseudobiospecific" are based on biomimetic approach, by which either simpler biomolecules (such as amino acids) or nonbiological molecules (such as tagged dyes, and hydrophobic side chains) can be employed significantly with higher chemical and physical stability for high-resolution affinity membrane-based analyte separations. The approach of utilizing synthetic ligands is based on the complementarity between structural features of ligand and the analyte, whereas pseudobiospecific ligands tend to have certain similarities, resulting in biomimetic properties for high-end biological applications. For example, electrospun nanofibers were functionalized with the common textile dye, cibacron blue, and used for affinity-based separation of albumin from blood samples and also for purification of dehydrogenase enzymes thanks to its site-specific binding ability [38, 39]. This suggested that the complementarily selection of suitable dyes enables highly efficient separations. Molecule-specific dyes usually possess improved stability and combined with low costs and precise binding coefficient make them an attractive solution for commercial separation.

Synthetic ligands also include immobilized metal chelates (IMC) and amino acids. Affinity separation based on IMC is generally employed for protein fractionation. IMC ligands are the most specific to enzymes and proteins, as affinity binding is generally based on its exposed surface containing amino acid residues. Some examples include iminodiacetate copper (II) complex (IDA-Cu(II)) and the nitrilotriacetate-nickel (NTANi-(II)) ligand, used for recombinant protein purification, performed with histidine residues. There are also review reports available on efficient separation of complex protein mixtures based on IMC [55–57]. Synthetic peptides derived from combinatorial protein libraries are also known for affinity separation of significant analytes such as human insulin and immunoglobulin (IgG). Synthetic peptide ligands are also known for immobilization on monoliths or molecularly imprinted polymers (MIPs). For instance, pentadeca peptides and hexadeca peptides have been used for purifying polyclonal antibodies [58]. Amino acids such as tryptophan, phenylalanine, and histidine are also employed as immunoadsorbers on polyvinyl alcohol substrate and display potential for next generation immobilized amino acid-based electrospun nanofibers, for protein and enzyme purifications.

Biological Ligands

Affinity membranes based on immobilized biological ligands, perform separation by utilizing the biological specific recognition capacity. In general, different ranges of biomolecules are used as affinity ligands; however, variation mostly exists in the method of immobilization adopted for different molecules. Sometimes, the same biomolecules render variation in the specificity with respect to difference in immobilization method. Widely employed ligands include covalently coupled protein A/G for purification of immunoglobulins, plasma, serum, or any cell culture [45]. Although biological ligands are well known for their remarkably greater specificity, limitations do exist with poor thermal and chemical stability and complex immobilization procedures. Another major problem deals with a probable leaching of the affinity ligand, which can lead to final product contamination. Biological ligands can be either biomolecule-specific or group-specific, which means that the biological recognition parameters are confined to a group of molecules rather than targeting a specific conformation of the ligate. Most of the antibodies fall under specific biological ligands whereas group-specific ligands include larger proteins, coenzymes, and receptor molecules. For instance, several fatty acids, metabolic toxins, and drugs usually bind preferentially with human albumin fraction. Even amino acids like lysine have been widely used for enzyme and protein purification [59, 60], as mentioned in Table 8.2.

Table 8.2 Different activation methods and immobilization ligands for developing affinity membranes for filtration applications

Application	Membrane	Activation/Immobilization	Reference
Bilirubin removal	Poly(vinyl alcohol-co-ethylene) (PVA-co-PE) nanofibrous	Sodium hydroxide and cyanuric chloride/diamine	[56]
	Polyamide	Epibromohydrin/polylysine	[36]
	Chitosan coated polyamide	Bisoxirane/lysine	[55]
	Aluminum oxide -silica composite membrane	Glutaraldehyde/lysine	[36]
Recombinant protein purification	Cellulose diacetate	Surfactant—like ligands	[57]
Lysozyme purification	Polyamide	Formaldehyde and Bisoxirane/ hydroxyethyl cellulose	[58]
IgG purification	Polysulfone	Diamine/hydroxyethyl cellulose	[59]
Protein purification	Polysulfone ultrafine fibers	Carboxyl groups/toluidine blue O dye	[3, 4]
Bovine globulin separation	Polyethylene	Amino acid immobilization	[60]
Bromelain removal	Polyacrylonitrile	Glutaraldehyde/cibacron blue dye	[33, 34]
Albumin adsorption	Polysulfone	Diamino-dipropylamine/ cibacron blue dye	[4]
Plasminogen purification	Regenerated cellulose	Epoxy/lysine	[54]
IgG separation	Dextran and polyvinyl alcohol coated polyamide	Formaldehyde and bisoxirane/protein A	[66]

Molecularly Imprinted Polymers

Molecularly imprinting polymers (MIPs) offer a real opportunity to deliver the required size and recognition sites due to site-specific ligands incorporated in the polymer matrix by molecular imprinting technology [19]. In this entirely different technology, the functional and cross-linking monomers get polymerized to form the molecular template along with target molecules intended for separation—the target components are then separated during the application. As the ligand binding is being introduced complementary to the polymer shape and functionality (Fig. 8.3), specific interaction sites will develop analogous to that of targeting antibodies [67, 68]. This method of binding enables improved efficiency of separation and extends the applications in terms of therapeutic systems. An additional advantage is also imparted as the cross-linked MIPs are both chemically and physically stable compared to bulk polymers.

Affinity separation based on MIPs would enhance the site-specific binding and purification/isolation efficiency. Hence, ES combined with this molecular imprinting

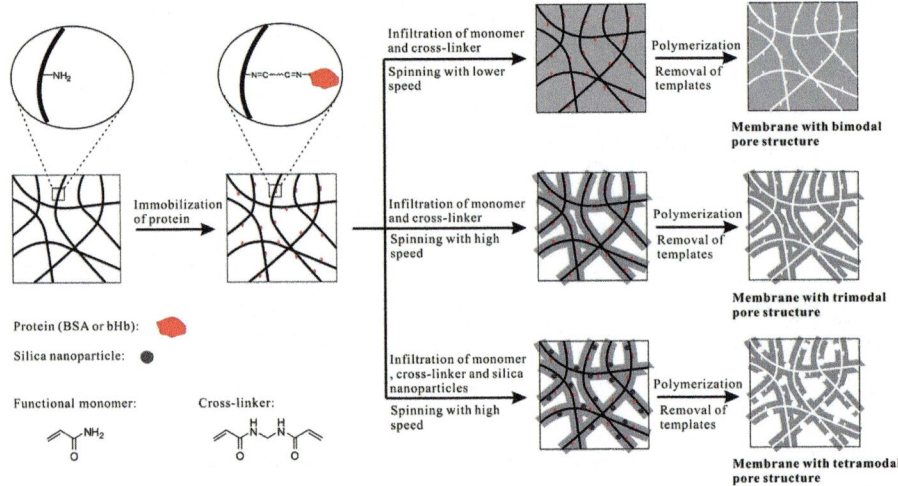

Fig. 8.3 Schematic representation of preparation of molecularly imprinted affinity membranes. Reprinted with permission from Royal Society of Chemistry [16]

technology enables fabrication of molecularly imprinted nanofiber membranes, which is gaining interest in the field of separation, immunoassays, biosensors, and self-healing reactions. Some of the nanostructured MIPs also can be employed for chiral and industrial filtration, benefitting from the controlled pore geometries and target recognition sites resulting from MIPs. Since the initial research on MIPs, efforts have been focused on synthesizing various physical geometries of MIP for desired end applications. Present applications focus mainly on the removal of specific protein molecules, especially peptide chains and amino acids.

8.6 Applications of Electrospun Affinity Membranes

Most commonly employed applications of affinity membranes use immobilized monoclonal antibodies as biological ligands for immunoaffinity separation. As antibodies are highly molecule-specific, the potential of affinity membranes lies in a wide range of biotechnological applications and also paves the way for new challenges in clinical therapies and diagnostics. With the advent of ES for affinity nanofiber synthesis, novel applications related to enzymatic and immunoassays were also performed using affinity membranes highlighting the significance in bio-analytical applications [69–71]. Electrospun affinity membranes immobilized with specific protein and antibodies as ligands have been utilized for immunoassays and microbial cell capture, respectively [72]. Small cartridges with membrane discs offer a method to revolutionize the field of sample preparation and extraction, for

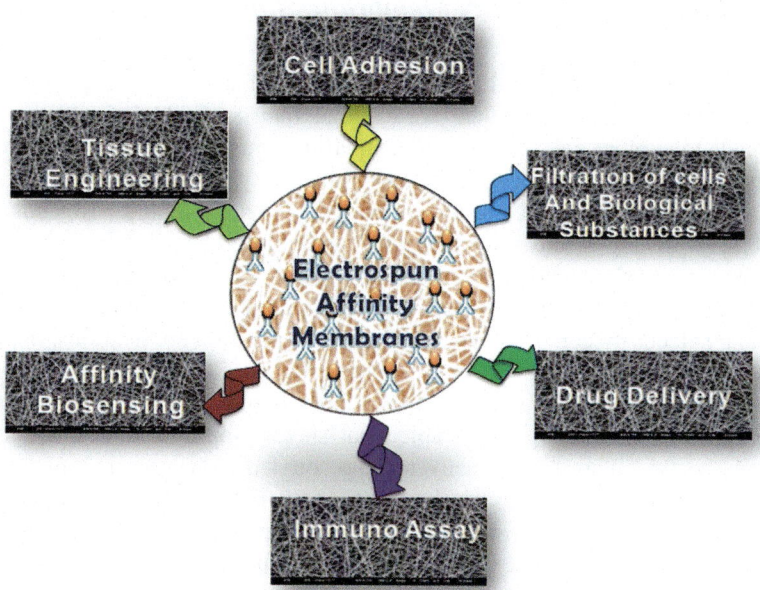

Fig. 8.4 Various applications based on electrospun affinity membranes

example, in immune extraction [52, 73]. Various biological applications of electrospun affinity membrane are schematically represented in Fig. 8.4.

Affinity-based bio-sensing applications are also expanding owing to the benefit of integrating electrospun membrane-based immunoassay system with microfluidic chip and signal sensing systems. In line with other advancements in medical devices, affinity membranes are finding applications in extracorporeal circuits, for the removal of toxic metabolites, antibodies, and substances such as in hemodialysis and peritoneal dialysis. The removal of amyloid components of human serum from the extracorporeal system circulating whole rat blood had been reported earlier in the past two decades [74] and applications related to immune-affinity membranes are also emerging [75]. Similarly, autoantibodies were isolated from human plasma, using ligand bound affinity membrane in extracorporeal circuits [76]. Applications pertaining to end product purification and polishing, biosensors, self-healing, and stimuli-responsive systems would also be practically viable, owing to continuing advances in the ES method incorporating varied design parameters and functionalities. Recent advancements are now emerging in both novel techniques and substrates including nanowires, micropillar, magnetic bead, mats, scaffolds, and microfluidic channels for translation towards final application. Despite several prospective applications of electrospun affinity membranes, the present chapter highlights applications related to cell and biomolecules filtration.

8.6.1 Electrospun Affinity Membranes for Cell Capture

Inspired by the comparable dimensions between electrospun membranes pores, cellular surface components, and extracellular matrix (ECM), significant efforts were made to explore the use of electrospun affinity membranes to capture cells for various disease identification [21]. Electrospun affinity membranes with high selectivity are prepared by surface modification using specific ligands. Ligands can be immobilized on the electrospun fibers using multiple chemical methods such as covalent, non-covalent, and coordination interactions. The choice of immobilization process helps to improve the interaction of cells with the membrane surface and thus molecular selectivity can be achieved in various separation applications. In general, reactive dyes, proteins, lectins, ion-exchange ligands, antigen–antibody ligands, and enzymes are employed to create affinity-based membranes. The affinity membranes are either single or stapled flat sheets, spiral wound, cassettes, and hollow fibers which are surface activated and coupled with appropriate ligands.

Affinity electrospun membranes have huge potential in capturing and identifying various tumor cells at early stages [61, 77]. Electrospun affinity membranes can selectively capture and detect very low numbers of tumor cells in blood streams due to their high surface area and availability of a great number of cell capturing sites. Recently, poly(lactic acid-*co*-glycolic acid) nanofibers modified with polyethyleneimine and hyaluronic acids were integrated onto a microfluidic platform to capture and detect circulating tumor cells (CTCs) from peripheral blood streams [77]. Compared to conventional CTC detection methods, electrospun affinity membranes selectively captured cancer cells at very low densities, with a reduced number of steps in the detection process. Wang et al. [61] investigated electrospun polystyrene (PS) fibers with incorporated antibodies to detect cancer markers, alpha fetoprotein (AFP), carcinoembryonic antigen (CEA), and vascular endothelial growth factors (VEGF). The electrospun substrate was shown to be more efficient and sensitive in detecting the cancer markers compared to conventional planar PS substrates. The improved performance was attributed to the much higher surface area of the nanofibers within the electrospun substrate.

In recent years, a great deal of interest has arisen in the use of hydrogels to fabricate affinity membranes to capture various proteins, due to their ability to form three-dimensional network, flexibility of chains, and water retention properties [78, 79]. Cibacron blue immobilized core-shell hydrogel affinity membranes based on nylon 6-chtosan/PVA were fabricated for separation of bovine serum albumin (BSA) [4]. By using hydrogel-based affinity electrospun membranes, the adsorption of BSA was significantly improved compared to traditional methods [3, 4, 80, 81]. Pre-spinning and post-spinning modifications of electrospun fibers show huge potential for next generation affinity membranes. Poly(styrene-co-maleimide) electrospun fibers were surface and bulk modified to successfully capture *Mycobacterium bovis* bacillus calmette-guerin (BCG) cells [82]. The authors noted that the agents used for the modification of the fibers played a crucial role in the capture of BCG cells. Recently, Lan et al. [83] fabricated reusable electrospun affinity membranes made of functionalized cellulose diacetate (CDA) with various func-

Table 8.3 Electrospun polymeric affinity membranes in biological applications

Polymer/functionalization	Application	Reference
Isocyanate functionalized electrospun poly(ether-urethane-urea) nanofiber	IgG purification	[77]
Polyethersulfone functionalized with cibacron blue F3GA (CB)	Dynamic adsorption of bovine serum albumin (BSA)	[78]
Regenerated cellulose nanofiber functionalized with proteins	IgG purification	[79]
Cellulose surface functionalized with Cibacron blue F3GA (CB)	Biomolecules separation	[33]
Hybrid chitosan/nylon-6	Papain purification	[33, 34]
Poly(styrene-co-maleic anhydride)	Microorganisms BCG	[75]
Glutaraldehyde functionalized Polyacrylonitrile nanofiber	Bromelain adsorption	[33, 34]
Polysulfone ultra-fine fiber surface functionalized with Cibacron blue F3GA	Albumin adsorption	[4]
Chitosan functionalized nylon-6	Dye affinity	[80]
Polydopamine functionalized doxorubicin loaded poly(L-lactide)	Antitumor effect	[81]
Hybrid collagen and Poly(lactide-co-ε-caprolactone)	Tissue engineering	[82]
Nylon 6/zinc doped hydroxyapatite membrane	Protein separation	[83]
Polyacrylonitrile functionalized with dual-vinyl and tri-vinyl monomers	Bacteria and viruses removal	[84]
Zinc doped hydroxyapatite containing nylon 6	Protein adsorption	[85]
Surface glycosylated poly(acrylonitrile-co-hydroxyethyl methacrylate)	Lectin adsorption	[86]
Heat and alkali treated cellulose functionalized with Cibacron blue F3GA	Biomolecules	[31]
Polyvinylalcohol	Papain	[87]

tional groups such as $-OH$, $-ONO_2$, and $-CH_2COCH_3$, for the separation and recovery of BSA. The membranes, fabricated with modified CDA, displayed maximum adsorption capacity for BSA (300.11 mg/g) compared to unmodified CDA (18.63 mg/g). There has been significant research endeavors devoted to the study of the nanoscale interactions between live cells and various substrate materials; however, it is crucial to understand the benefit of interactions between nanoscale features of nanofibers embedded in cellular surface components (e.g., microvilli and filopodia) and ECM scaffolds.

8.6.2 Affinity Membranes in Biological Applications

In conventional affinity separation, the configuration of flat sheet, hollow fiber discs has been widely employed for the separation of proteins, antibodies, immunoglobulins, and pristine amino acids. A selection of publications on affinity membranes for biomolecules isolation and purification is summarized in Table 8.3.

PAN-based electrospun membranes containing cyclic sugars were prepared to recover and purify mixtures of proteins [95]. This isolation is based on selective recognition between saccharide residues and proteins, which is the first step in a host of additional cell–cell interactions, such as blood coagulation, immune response, viral infection, inflammation, and embryogenesis, and cellular signal transfer [3, 4]). As a result, the teams developed a series of affinity nanofibrous membranes to capture specific target molecules. On the other hand, methacrylic acid (MAA) was grafted to the nanofiber surface to capture Toluidine Blue O (TBO) molecules ionically and BSA chemically [3]. Analysis by confocal laser scanning microscope showed that the labeled BSA covered the electrospun membrane surface uniformly and was distributed evenly into the fibrous membranes. These electrospun membranes showed high water permeability, good reusability, and the capacity to capture target molecules. More recently, cellulose nanofibrous composite membranes made up of electrospun cellulose nanofabric resulted in the rejection of polymeric beads (as small as 50 nm), indicating that these membranes have potential application in virus removal [96]. Surface functionalization of microfibrous poly(ε-caprolactone) (PCL) with surface covalently immobilized antibodies has been reported to develop electrospun nanofibrous membranes for immuno-enhanced tissue scaffolds [25].

8.7 Future Outlook and Challenges

Affinity membranes for cell capturing and removal of biological substances have gained practical viability, whereas therapeutic and biomedical applications of affinity membranes remain in their infancy. Electrospun nanofibers hosting nanomaterial carriers have gained widespread attention over the last decade. Efforts to ensure improved performance of nanofibers for biological applications remain of major interest, requiring a tailored material response for site-specific interactions or drug loading for targeted and controlled delivery applications [97]. The ES technique for developing conventional polymeric fibers or affinity nanofibers is considered commercially viable at laboratory scale. Yet, the transition from laboratory to large-scale commercialization requires disruptive advancement in the current ES technology and ligand binding methods. Moreover, growing demands for filtration/separation materials in this modern era could be met by exploring nonconventional technologies to move the versatile, multifunctional ES-based affinity membrane technology towards specialized, industrial end-applications, not limited to biomolecule separation or cell capture alone.

8.8 Conclusions

Continuous growing trends in biotechnological, pharmaceutical, and biomedical field demand economic but efficient methods for the separation of a wide variety of biomolecules. One viable commercial route for driving the affinity membrane separation into large-scale production is by incorporating sophisticated design and functionalities via the versatile ES membrane fabrication technology. At the same time, the increasing multifunctional nanomaterials and modification strategies bring critical challenges in the preparation of electrospun affinity membranes which need to match the best performance criteria. Research reports relating to parametric analysis on improving the ES process and also the impact of conjugating polymer modification methods would offer a scalable, economical method of producing affinity functionalized nanofibers. Many multifunctional and composite membranes have been developed over recent years, but there is still a need to develop biocompatible mats using materials from renewable sources for recovery of cells and biologically active substances. The use of natural antimicrobials and biodegradable polymers would have to prove competitive with approaches that are more conventional.

References

1. Klein E (2000) Affinity membranes: a 10-year review. J Membr Sci 179:1–27
2. Thommes J, Kula MR (1995) Membrane chromatography-an integrative concept in the downstream processing of proteins. Biotechnol Prog 11:357–367
3. Ma Z, Kotaki M, Ramakrishna S (2006) Surface modified nonwoven polysulphone (PSU) fiber mesh by electrospinning: a novel affinity membrane. J Membr Sci 272:179–187
4. Ma Z, Masaya K, Ramakrishna S (2006) Immobilization of Cibacron blue F3GA on electrospun polysulphone ultra-fine fiber surfaces towards developing an affinity membrane for albumin adsorption. J Membr Sci 282:237–244
5. Gao H, Sun X, Gao C (2017) Antifouling polysulfone ultrafiltration membranes with sulfobetaine polyimides as novel additive for the enhancement of both water flux and protein rejection. J Membr Sci 542:81–90
6. Haider S, Park SY (2009) Preparation of the electrospun chitosan nanofibers and their applications to the adsorption of Cu (II) and Pb (II) ions from an aqueous solution. J Membr Sci 328:90–96
7. Zhang H, Nie H, Li S et al (2008) Electrospun nylon nanofiber as affinity membrane for papain adsorption. J Biotech 136:S416
8. Ma Z, Mao Z, Gao C (2007) Surface modification and property analysis of biomedical polymers used for tissue engineering. Colloids Surf B 60:137–157
9. Karim Z, Mathew AP, Kokol V et al (2016) High-flux affinity membranes based on cellulose nanocomposites for removal of heavy metal ions from industrial effluents. RSC Adv 6:20644–20653
10. Zhang H, Jia X, Han F et al (2013) Dual-delivery of VEGF and PDGF by double-layered electrospun membranes for blood vessel regeneration. Biomaterials 34:2202–2212
11. Web of Sciences (2017) https://apps.webofknowledge.com/. Accessed 06 Dec 2017

12. Ramakrishna S, Fujihara K, Teo WE et al (2005) An introduction to electrospinning and nanofibers chapter 2. World Scientific Publishing, Singapore
13. Nasreen SAAN, Sundarrajan S, Nizar SAA et al (2013) Advancement in electrospun nanofibrous membranes modification and their application in water treatment. Membranes 3(4):266–284
14. Alam H, Ramakrishna S (2012) A review on the enhancement of figure of merit from bulk to nano-thermoelectric materials. Nano Energy 2:190–212
15. Raghavan P et al (2012) Electrospun polymer nanofibers: the booming cutting edge technology. React Funct Polym 72(12):915–930
16. Shin SH, Purevdorj O, Planell JA et al (2012) A short review: recent advances in electrospinning for bone tissue regeneration. J Tissue Eng 3(1):2041731412443530. https://doi.org/10.1177/2041731412443530
17. Teo WE, Inai R, Ramakrishna S (2011) Technological advances in electrospinning of nanofibers. Sci Technol Adv Mater 12(1):013002
18. Wang X, Hsiao BS (2016) Electrospun nanofiber membranes. Curr Opin Chem Eng 12:62–81
19. Zhu T, Xu D, Wu Y et al (2013) Surface molecularly imprinted electrospun affinity membranes with multimodal pore structures for efficient separation of proteins. J Mater Chem B 1:6449–6458
20. Nada AA, James R, Shelke NB et al (2014) A smart methodology to fabricate electrospun chitosan nanofiber matrices for regenerative engineering applications. Polym Adv Technol 25:507–515
21. Son YJ, Kang J, Kim HS et al (2016) Electrospun nanofibrous sheets for selective cell capturing in continuous flow in microchannels. Biomacromolecules 17:1067–1074
22. Wang X, Um IC, Fang D et al (2005) Formation of water-resistant hyaluronic acid nanofibers by blowing-assisted electro-spinning and non-toxic post treatments. Polymer 4:4853–4867
23. Hu X, Liu S, Zhou G (2014) Electrospinning of polymeric nanofibers for drug delivery applications. J Control Release 185:12–21
24. Chakraborty S, Liao IC, Adler A et al (2009) Electrohydrodynamics: a facile technique to fabricate drug delivery systems. Adv Drug Deliv Rev 61:1043–1054
25. Guex AG, Hegemann D, Giraud MN et al (2014) Covalent immobilisation of VEGF on plasma-coated electrospun scaffolds for tissue engineering applications. Colloids Surf B Biointerfaces 123:724–733
26. Tonglairoum P, Ngawhirunpat T, Rojanarata T et al (2015) Fabrication of a novel scaffold of clotrimazole-microemulsioncontaining nanofibers using an electrospinning process for oral candidiasis applications. Colloids Surf B Biointerfaces 126:18–25
27. Zhang M, Wang Z, Wang Z et al (2011) Immobilization of anti-CD31 antibody on electrospun poly(e-caprolactone) scaffolds through hydrophobins for specific adhesion of endothelial cells. Colloids Surf B Biointerfaces 85:32–39
28. Chen Z, Chen Z, Zhang A et al (2016) Electrospun nanofibers for cancer diagnosis and therapy. Biomater Sci 4:922–932
29. Chung HY, John RB, Gogins MA et al (2008) Use of crosslinked polyvinyl alcohol in polymers with improved environmental stability. EP1925352A1, 3–15
30. Ebara M, Kotsuchibashi Y, Uto K et al (2014) Smart hydrogels In: Smart biomaterials, pp 1–373
31. Shabafrooz V, Mozafari M, Vashaee D et al (2014) Electrospun nanofibers: from filtration membranes to highly specialized tissue engineering scaffolds. J Nanosci Nanotechnol 14:522–534
32. Li H, Li C, Zhang C et al (2014) Well dispersed copper nanorods grown on the surface functionalized PAN fibers and its antibacterial activity. J Appl Polym Sci 131:41011
33. Li X, Wang C, Yang Y (2014) Dual-biomimetic superhydrophobic electrospun polystyrene nanofibrous membranes for membrane distillation. ACS Appl Mater Interfaces 6:2423–2430

34. Yang Z, Si J, Cui Z et al (2017) Biomimetic composite scaffolds based on surface modification of polydopamine on electrospun poly(lactic acid)/cellulose nanofibrils. Carbohydr Polym 174(15):750–759
35. Ma Z, Kotaki M, Yong T, He W, Ramakrishna S (2005a) Surface engineering of electrospun polyethylene terephthalate (PET) nanofibers towards development of a new material for blood vessel engineering. Biomaterials 26(15):2527–2536
36. Ma Z, Kotaki M, Ramakrishna S (2005b) Electrospun cellulose nanofiber as affinity membrane. J Membr Sci 265:115–123
37. Yoshikawa M, Ooi T, Izumi J (1999) Alternative molecularly imprinted membranes from a derivative of natural polymer, cellulose acetate. J Appl Polym Sci 72:493–499
38. Zhang H, Nie H, Yu D et al (2010) Surface modification of electrospun polyacrylonitrile nanofiber towards developing an affinity membrane for bromelain adsorption. Desalination 256:141–147
39. Zhang H, Wu C, Zhang Y et al (2010) Elaboration, characterization and study of a novel affinity membrane made from electrospun hybrid chitosan/nylon-6 nanofibers for papain purification. J Mater Sci 45:2296–2304
40. Reddy PS, Kobayashi T, Fujii N (2002) Recognition characteristics of dibenzofuran by molecularly imprinted polymers made of common polymers. Eur Polym J 38:779–785
41. Shi W, Zhang F, Zhang G et al (2010) Polylysine-immobilized affinity nylon membrane used for bilirubin adsorption. Mol Simul 29:787–790
42. Trotta F, Drioli E, Baggiani C, Lacopo D (2002) Molecularly imprinted polymeric membrane for naringin recognition. J Membr Sci 201:77–84
43. Ulbricht M, Malaisamy R (2005) Insights into the mechanism of molecular imprinting by immersion precipitation phase inversion of polymer blends via a detailed morphology analysis of porous membranes. J Mater Chem 15:1487–1497
44. Ulbricht M (2006) Advanced functional polymer membranes. Polymer 47:2217–2262
45. Avramescu MME, Wessling M, Borneman Z (2015) In: Pabby AK, Rizvi SSH, Requena AMS (eds) Membrane and monolithic convective chromatographic supports, 2nd edn. Taylor & Francis Group, Burlington, p 42
46. Hage DS, Cazes J (eds) (2006) Handbook of affinity chromatography, 2nd edn. CRC Press, Boca Raton
47. Fleschin S, Bunaciu AA, Scripcariu M (2004) Preferable methods for immobilized biocatalysts in enzyme electrode construction. Rom Biotechnol Lett 9:1947–1958
48. Suen SY, Lin SY, Chiu HS (2000) Effects of spacer arms on Cibacron blue 3GA immobilization and lysozyme adsorption using regenerated cellulose membrane discs. Ind Eng Chem Res 39:478–487
49. Acikara OB, Citoglu GS, Ozbilgin S, Ergene B (2013) Affinity chromatography and importance in drug discovery. In Tech open access. doi: https://doi.org/10.5772/55781
50. Ayyar BV, Arora S, Murphy C, O'Kennedy R (2012) Affinity chromatography as a tool for antibody purification. Methods 56:116–129
51. Zou H, Luo Q, Zhou D (2001) Affinity membrane chromatography for the analysis and purification of proteins. J Biochem Biophys Methods 49:199–240
52. Quiros J, Boltes K, Rosal R (2016) Bioactive applications for electrospun fibers. Polym Rev 56(4):631–667
53. Chakrabarty T, Kumar M, Rajesh KP et al (2010) Nanofibrous sulfonated poly (ether ether ketone) membrane for selective electro-transport of ions. Sep Purif Technol 75:174–182
54. Wang J, Kang Q, Lv X et al (2013) Simple patterned Nanofiber scaffolds and its enhanced performance in immunoassay. PLoS One 8(12):82–88
55. Block H, Maertens B, Spriestersbach A, Brinker N, Kubicek J, Fabis R, Labahn J, Schafer F (2009) Immobilized-metal affinity chromatography (IMAC): a review, methods in enzymology, vol 463. Elsevier, Amsterdam, pp 439–473

56. De Aquino LC, De Sousa HR, Miranda EA, Vilela L, Bueno SM (2006) Evaluation of IDA-PEVA hollow fiber membrane metal ion affinity chromatography for purification of a histidine-tagged human proinsulin. J Chromatogr B Analyt Technol Biomed Life Sci 834(1–2):68–76

57. Serpa G, Augusto EFP, Tamashiro WMSC, Ribeiro MB, Miranda EA, Bueno SMA (2005) Evaluation of immobilized metal membrane affinity chromatography for purification of an immunoglobulin G1 monoclonal antibody. J Chromatogr B 816:259–268

58. Taguchi H, Sunayama H, Takano E, Kitayama Y, Takeuchi T (2015) Preparation of molecularly imprinted polymers for the recognition of proteins via the generation of peptide-fragment binding sites by semicovalent imprinting and enzymatic digestion. Analyst 140:1448–1452

59. Boi C, Castro C, Sarti GC (2015) Plasminogen purification from serum through affinity membranes. J Membr Sci 475:71–79

60. Shi W, Zhanga F, Zhanga G (2005) Adsorption of bilirubin with polylysine carrying chitosan-coated nylon affinity membranes. J Chromatogr B 819:301–306

61. Wang W, Zhang H, Zhang Z et al (2017) Amine-functionalized PVA-co-PE nanofibrous membrane as affinity membrane with high adsorption capacity for bilirubin. Colloids Surf B 150:271–278

62. Honjo T, Hoe K, Tabayashi S et al (2013) Preparation of affinity membranes using thermally induced phase separation for one-step purification of recombinant proteins. Anal Biochem 434:269–274

63. Beeskow TC, Kusharyoto W, Anspach FB et al (1995) Surface modification of microporous polyamide membranes with hydroxyethyl cellulose and their application as affinity membranes. J Chromatogr A 715:49–65

64. Klein E, Eichholz E, Yeager DH (1994) Affinity membranes prepared from hydrophilic coatings on microporous polysulfone hollow fibers. J Membr Sci 90:69–80

65. Kim M, Saito K, Furusaki S et al (1991) Adsorption and elution of bovine γ-globulin using an affinity membrane containing hydrophobic amino acids as ligands. J Chromatogr A 585:45–51

66. Castilho LR, Deckwer WD, Anspach FB (2000) Influence of matrix activation and polymer coating on the purification of human IgG with protein A affinity membranes. J Membr Sci 172:269–277

67. Tokonami S, Shiigi H, Nagaoka T (2009) Review: micro- and nanosized molecularly imprinted polymers for high-throughput analytical applications. Anal Chim Acta 641:7–13

68. Bompart M, Haupt K, Ayela C (2012) Micro and nanofabrication of molecularly imprinted polymers. Top Curr Chem 325:83–110

69. Lee Y, Lee JH, Son KJ, Koh WG (2011) Fabrication of hydrogel-micropatterned nanofibers for highly sensitive microarray-based immunosensors having additional enzyme-based sensing capability. J Mater Chem 21:4476–4483

70. Liu Y, Yang D, Yu T, Jiang X (2009) Incorporation of electrospun nanofibrous PVDF membranes into a microfluidic chip assembled by PDMS and scotch tape for immunoassays. Electrophoresis 30:3269–3275

71. Yang D, Niu X, Liu Y, Wang Y, Gu X, Song L, Zhao R, Ma L, Shao Y, Jiang X (2008) Electrospun nanofibrous membranes: a novel solidsubstrate for microfluidic immunoassays for HIV. Adv Mater 20:4770–4775

72. Senecal KJ, Senecal AG, Pivarnik PE, Mello CM, Soares JW, Schreuder-Gibson HL (2010) Electrospun nanofibrous membrane assembly for use in capturing chemical and/or biological analytes, US Patent, US20100240121 A1, 1–7

73. Zha Z, Cohn C, Dai Z et al (2011) Nanofibrous lipid membranes capable of functionally immobilizing antibodies and capturing specific cells. Adv Mater 23:3435–3440

74. Adachi T, Mogi M, Harada M, Kojima K (1996) Selective removal of human serum amyloid P component from rat blood by use of an immunoaffinity membrane in an extracorporeal circulation system. J Chromatogr B Biomed Appl 682(1):47–54

75. Bereli N, Yavuz H, Denizli A (2015) Immunoaffinity membranes. In: Drioli E, Giorno L (eds) Living reference work entry encyclopedia of membranes. Springer, Berlin, pp 1–2

76. Haupt K, Bueno SMA (2009) In: Wilson ID, Adlard ER, Cooke M, Poole CF (eds) Affinity membranes, handbook of methods and instrumentation in separation science. Academic Press, San Diego
77. Xu G, Tan Y, Xu T et al (2017) Hyaluronic acid-functionalized electrospun PLGA nanofibers embedded in a microfluidic chip for cancer cell capture and culture. Biomater Sci 5:752–761
78. Bell CL, Peppas NA (1995) Biomedical membranes from hydrogels and interpolymer complexes. Adv Polym Sci 22:125–176
79. Mondal S, Li C, Wang K (2015) Bovine serum albumin adsorption on Glutaraldehyde cross-linked chitosan hydrogels. J Chem Eng Data 60:2356–2362
80. Luong JHT, Nguyen AL (1992) In: Tsao GT (ed) Bioseparation, vol 47. Springer, Berlin, pp 137–158
81. Tong U, Xu X, Wang H et al (2015) Solution-blown core–shell hydrogel nano fibers forbovine serum albumin affinity adsorption. RSC Adv 5:83232
82. Cronje L, Klumperman B (2013) Modified electrospun polymer nanofibers as affinity membranes: the effect of pre-spinning modification versus post-spinning modification. Eur Polym J 49:3814–3824
83. Lan T, Shao Z, Gu M et al (2015) Electrospun nanofibrous cellulose diacetate nitrate membrane for protein separation. J Membr Sci 489:204–2011
84. Bamford CH, Al-Lamee KG, Purbrick MD, Wear TJ (1992) Studies of a novel membrane for affinity separations: functionalisation and protein coupling. J Chromatogr A 606:19–31
85. Ma Z, Lan Z, Matsuura T et al (2009) Electrospun polyethersulfone affinity membrane: membrane preparation and performance evaluation. J Chromatogr B 877:3686–3694
86. Ma K, Chan CK, Liao S et al (2008) Electrospun nanofiber scaffolds for rapid and rich capture of bone marrow-derived hematopoietic stem cells. Biomaterials 29:2096–2103
87. Zhang HT, Han J, Xue Y et al (2009) Surface modification of electrospun nylon nanofiber based dye affinity membrane and its application to papain adsorption. Bioinform Biomed Eng. In: 3rd International conference on bioinformatics and Biomedical engineering, 11–13 June 2009
88. Yuan Z, Zhao X, Wang X et al (2014) Promotion of initial anti-tumor effect via polydopamine modified doxorubicin-loaded electrospun fibrous membranes. Int J Clin Exp Pathol 7(9):5436–5449
89. Xiaomin H, Wei F, Bei F et al (2013) Electrospun collagen/Poly(L-lactic acid-co-ε-caprolactone) hybrid Nanofibrous membranes combining with sandwich construction model for cartilage tissue engineering. J Nanosci Nanotechnol 13:3818–3825
90. Esfahani H, Prabhakaran MP, Salahi E et al (2016) Electrospun nylon 6/zinc doped hydroxyapatite membrane for protein separation: mechanism of fouling and blocking model. Mater Sci Eng C 59:420–428
91. Ma H, Hsiao BS, Chu B (2014) Functionalized electrospun nanofibrous microfiltration membranes for removal of bacteria and viruses. J Membr Sci 452:446–452
92. Esfahani H, Prabhakaran MP, Salahi E et al (2015) Protein adsorption on electrospun zinc doped hydroxyapatite containing nylon 6 membrane: kinetics and isotherm. J Colloid Interface Sci 443:143–152
93. Che AF, Huang XJ, Xu ZK (2011) Polyacrylonitrile-based nanofibrous membrane with glycosylated surface for lectin affinity adsorption. J Membr Sci 366:272–277
94. Moreno-Cortez IE, Romero-García J, González-González V et al (2015) Encapsulation and immobilization of papain in electrospun nanofibrous membranes of PVA cross-linked with glutaraldehyde vapor. Mater Sci Eng C 52:306–314
95. Yang Q, Li JJ, Hu MX et al (2006) Nanofibrous sugar sticks via electrospinning. Macromol Rapid Comm 42:1942–1948
96. Huang W, Wang Y et al (2016) Fabrication of flexible self-standing all-cellulose nanofibrous composite membranes for virus removal. Carbohydr Polym 143:9–17

Chapter 9
Electrospun Nanofibre Filter Media: New Emergent Technologies and Market Perspectives

Ankita Poudyal, Gareth W. Beckermann, Naveen Ashok Chand, Iain C. Hosie, Adam Blake, and Bhuvaneswari Kannan

Abstract With the continual rise in the levels of urban and industrial pollution, the demand for high-efficiency electrospun nanofibre filters has never been greater. Many of the most widely recognised filter manufacturers in the world have already taken a lead in incorporating electrospun nanofibre materials into their filters. Now, the next phase of growth in this market is in the development of functional filters that selectively capture targets of interest, such as VOCs, PM2.5 particles, heavy metal ions, dyes, viruses and many other contaminants. This chapter is a summary of the emerging technologies that can be used in conjunction with electrospun nanofibres for the high-performance filtration of fine particles.

9.1 Introduction

The increasing frequency of red alert warnings due to high levels of air and water pollution is resulting in widespread disruption and forcing governments to enforce more effective health and safety mechanisms to protect the public. A huge spike in the occurrence of diseases such as asthma, cancer, high blood pressure, birth defects [1, 2] and many other cardiovascular and respiratory problems [3] in the past couple of decades is directly proportionate to the increased air and water pollution levels all over the world. Air and water are the bulk transportation media for the transmission of contaminants [4]. The pollutants found in both media consist of complex mixtures of gases and particles that could be biological or non-biological in nature. Polluting particles typically have hundreds of chemicals adsorbed onto their surfaces that could have mutagenic and carcinogenic effects on humans [1–3, 5]. Most of these

A. Poudyal
Kode Biotech, Auckland University of Technology, Auckland, New Zealand

G. W. Beckermann · N. A. Chand · I. C. Hosie · A. Blake · B. Kannan (✉)
Revolution Fibres Ltd, Auckland, New Zealand
e-mail: bhuvana@revolutionfibres.com

© Springer International Publishing AG, part of Springer Nature 2018
M. L. Focarete et al. (eds.), *Filtering Media by Electrospinning*,
https://doi.org/10.1007/978-3-319-78163-1_9

197

toxic particulate compounds are smaller than 1 micrometre (μm) in diameter. Conventional mechanical fibrous filters remove micrometre-sized particles with high efficiency. However, for particles in the submicron size range, electrospun media with fine nanofibres are considered to be better as they offer enhanced filtration performance due to their high surface area and small pore dimensions. The electrospinning process has been hugely discussed and explored by researchers elsewhere [6–10]. Lately, there has been much progress towards the modification of electrospun nanofibres to provide them with additional functionality that enhances their filtration efficiencies. Some recent research trends regarding modification of electrospun nanofibres have the potential to be translated to commercial applications [6], and the market insights of such technologies for filtration applications are highlighted in the forthcoming sections.

9.2   Nanotechnology in the Air Filtration Industry

The development of nanofibre filter media for air filtration dates back to 1939 when Nathalie D. Rozenblum and Igor V. Petryanov-Sokolov developed the Petryanov filter (PF) in Russia. Initially, the filters were used as smoke filter elements for gas masks, but after the Second World War, the Russians set up five enterprises to manufacture these filters to capture radioactive aerosols in nuclear power plants. Although the fibres showed excellent filtration properties, the products were confined to the USSR due to secrecy with the West. The first submicron fibre filtration media was commercialised in the USA by Donaldson Co. in the 1980s. Since then, the technique has been adopted by many industries and research groups and has undergone massive improvements to suit different filtration markets.

The increasing number of warnings from health organisations and agencies combined with increasingly stringent government regulations and health and safety concerns from the public is pushing the development of next-generation filter media with more efficient filtration capabilities. Simultaneously, this pollution-conscious climate has led to the emergence of new technological disrupters in air filtration industry.

Nanofibres have been identified as one such disrupting technology for existing filter industries that utilise non-woven microglass or macro-porous fabrics. The materials containing nanofibres are highly regarded in air filtration applications because of their high surface area which enables them to electrostatically attract dust and pollen particles, and because of their small pore sizes which effectively sieve fine hazardous particulates. Moreover, nanofibre fabrication processes often allow for the incorporation of functional molecules within the fibres to better capture target particulates, viruses and hazardous gases with greater efficiency.

The total market for nanofibre products was predicted to grow at a compound annual growth rate (CAGR) of 30.3% from 2012 to 2017, reaching global revenues of $570.2 million by 2017. Out of this, 74.7% of all revenues, or $425 million USD,

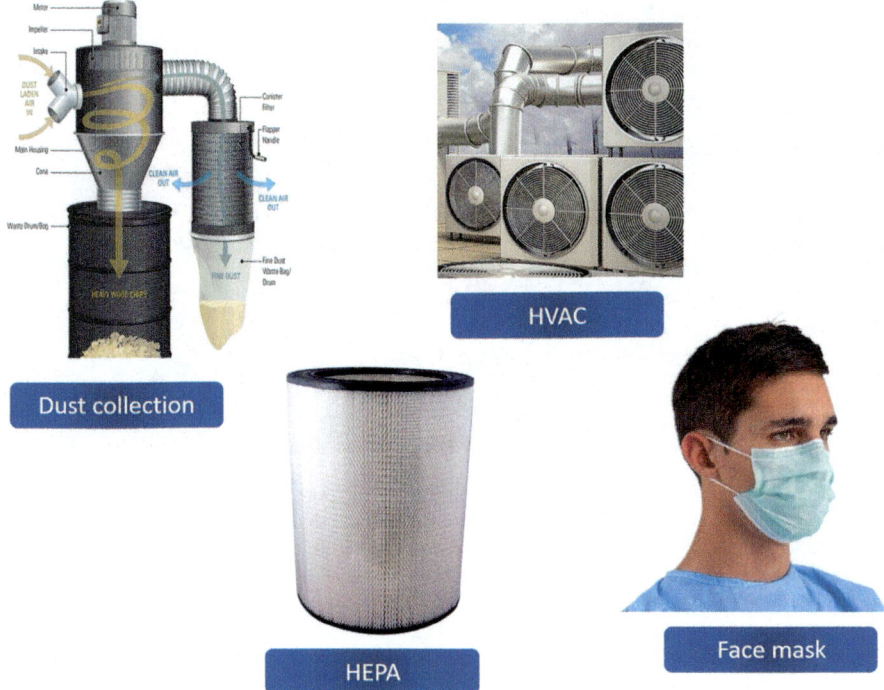

Fig. 9.1 Application of nanofibres in the air filtration industry

was associated primarily with the utilisation of nanofibres in the fabrication of filtration media [11].

Nanofibre filtration markets are generally focused on the sale of both indoor and outdoor products such as HEPA, HVAC and dust collection filters as well as personal face masks which include simple, surgical, N95, PM2.5 classification respiratory protection and high-functional face masks for military use (Fig. 9.1).

9.3   Market Drivers for Nanofibre Air Filtration Media

The biggest market driver for nanofibre air filters is the increasing awareness and concern over air pollution (Fig. 9.2).

Globally, governments are pledging to improve the existing air quality through new technologies. In 2012, 67% of China's energy consumption was fuelled by coal-burning power plants [12] and pollution emitted by these plants has been blamed for 257,000 premature deaths, 141 million sick days and 340,000 hospital visits in 2011 [13]. In regions that suffer from air pollution where the PM 2.5 concentration is high (that includes dust particles, volatile gases, virus and bacteria with

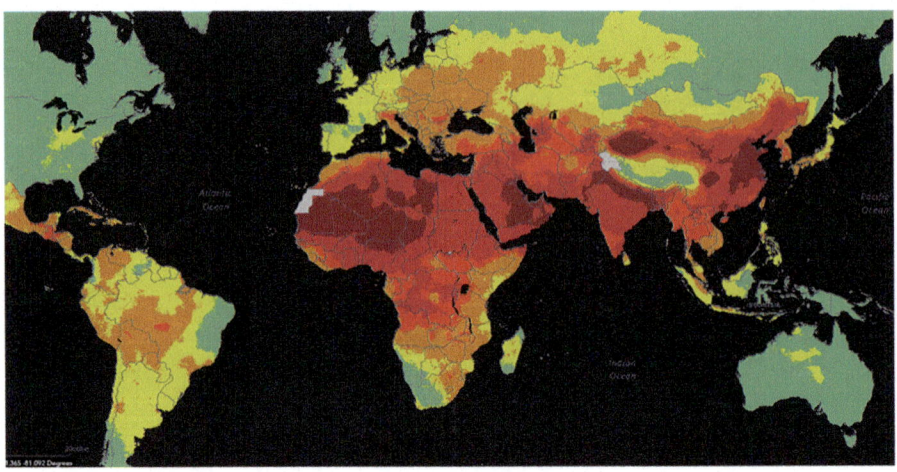

Fig. 9.2 World map indicating levels of air pollution. Large portions of the globe are coloured in yellow, orange, red, purple and green to indicate the magnitude of breaches of WHO air quality limits. Yellow/green indicates a low level of airborne pollution, and red indicates a high level of airborne pollution (reproduced from WHO website)

a diameter not larger than 2.5 μm), the public is often encouraged to take personal precautions and to wear a face mask while travelling. As a result of this, the Chinese market is one of the primary targets for face mask and residential air filtration system manufactures. China has invested around $277 billion to combat air pollution, primarily in areas affected by high levels of PM 2.5 pollutants [14]. China is the major consumer market for indoor air quality (IAQ), followed by the USA, UK and other Asian countries, and rapid uptake into other regions is expected over the next 5 years.

On the other hand, U.S. joint services continue to standardise air quality testing procedures and try to encourage the development and commercialisation of next-generation high filtration efficiency and low pressure drop particulate filtration technologies. The IAQ market in the USA was worth over $7.7 billion in 2013 and was expected to grow to $8.1 billion in 2014 and $11.4 billion by 2019 at a compound annual growth rate (CAGR) of 7.0% over 5 years [15]. According to the Environmental Protection Agency (EPA), most U.S. citizens spend nearly 90% of their time in indoor environments, such as homes, workplaces, classrooms and other school environments. The agency estimates that indoor levels of air pollutants can be two to five times higher, and occasionally 100-times higher than outdoor levels [15].

Although IAQ markets are driven by public awareness, the worsening of air pollution and its detrimental impact on health and well-being has driven strong growth in the consumer face mask market which includes PM 2.5 and surgical masks.

9.4   Air Filtration Media

Air pollutants are comprised of particles and gases such as Volatile Organic Compounds (VOCs), sulphur dioxides, nitrogen dioxides, ozone, viruses, dust, smoke, etc. Polycyclic aromatic hydrocarbons (PAHs) are other commonly occurring toxic airborne pollutants and are produced by the incomplete combustion of organic matter [16, 17].

Dust particles or pathogens typically occur as liquids or solids or combinations thereof. For example, dust particles and pathogens develop a liquid outer phase when they hit the surface of filtration media that has been humidified by atmospheric moisture or from the moisture from human respiration. An interfacial tension, therefore develops between the particle and the filter media.

These complexities of the composition of airborne pollutants have led to increases in demand for high efficiency filtration media consisting of multiple layers for capturing particulates of different sizes, forms and compositions. Unfortunately, multilayer filter media is often hindered by poor breathability and high pressure drop, which is undesirable for filters. Thus, researchers are putting in much effort to develop single-layer multifunctional nanofibre filtration media to address the issues of poor breathability and low filtration efficiency. It is also possible to utilise the inherently large surface areas of the nanofibres and to chemically modify and functionalise the nanofibre surfaces so that they are able to adsorb larger amounts and a wider range of pollutants than what would normally be possible. The use of nanofibres to filter different types of pollutants is discussed below.

9.4.1   Particulate Matter

Particulate matter (PM) is the sum of all solid and liquid particles that are suspended in air, many of which are hazardous. Smaller PMs, specifically PM2.5, consist of particles that are less than 2.5 μm in size and are small enough to penetrate deeply into the lung, irritate and corrode the alveolar wall and consequently impair lung function.

In general, the filtration of PMs occurs in two ways: surface filtration and depth filtration. Simply put, surface filtration occurs when particles are too large to fit or pass through the pores of a filter and are trapped on the surface of the filter media. The general mechanisms for the surface filtration of PMs include diffusion, interception, intermolecular interaction, straining, inertial impaction, gravitation and electrostatic interaction of particles on the filter surface (Fig. 9.3) [18].

Diffusion and interception can be the most important particle capture mechanisms of nanofibres. Diffusion results in the capture of fine particles below 0.5 mm on the fibre surface due to Brownian motion [19], whereas particle interception occurs when the distance between fibre surface and the particle centre is equal to or less than the particle radius [19, 20].

If not removed efficiently through these mechanisms, particulate contaminants tend to clog filter media and significantly reduce airflow. In addition to these

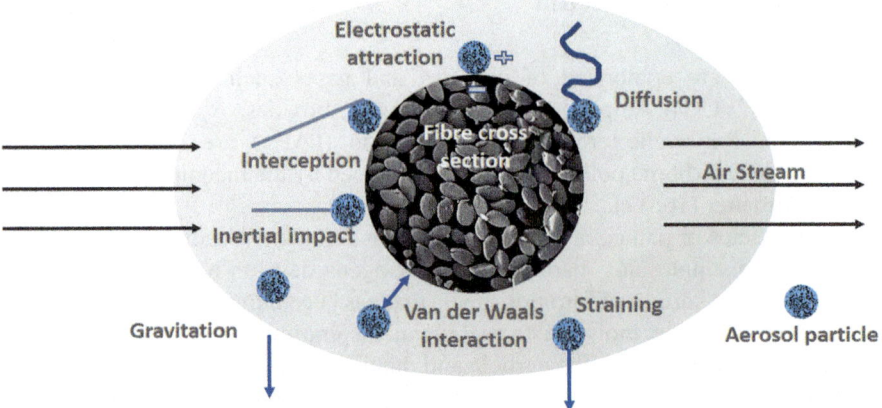

Fig. 9.3 Mechanisms of particulate matter filtration

attributes, nanofibre membranes exhibit strong van der Waals forces which enable them to attract submicron sized particles. The good interconnectivity of the pores also results in improvements in inertial impact and interception of particles. In addition, there is great potential to modify the nanofibre filtration media by incorporating additional functionality on the nanofibre surfaces which could enhance the chemical interaction and electrostatic attraction with particle contaminants.

On the other hand, depth filtration occurs when particles are small enough to fit into the pores of the filtration media but are trapped during their journey through the material. Consequently, the particle size distribution and the pore size of the filtration media determine whether or not surface or depth filtration occurs. Thus, in both surface filtration and depth filtration situations, the particles that are small enough to enter the filter structure are collected by chance interactions with the fibres of the filter. The selection of the ideal filter material for surface and depth filtration requires knowledge of the interfacial energies at the surface between the fibres and the particulates as well as surface interaction mechanisms.

Many other variables influence the efficiency of filter media in capturing particulates, including duct size, fan strength, air velocity and environmental conditions such as humidity.

In case of nanofibre air filters, as aforementioned, dust particles and airborne pathogens can have relatively wet surfaces when compared to the dry nanofibre material. Thus, when airborne pollutants touch the surface of the nanofibre, an interfacial tension arises at the solid–liquid or solid–solid boundary/interface from the polarity disparity of the solid phase of the fibre and moisture-containing pollutant particles. Due to this interfacial tension, the adherence between solid nanofibres and relatively wet particles is increased.

It is possible to enhance such interactions and adherence between the nanofibres and the particulates by introducing polar molecules to the nanofibre polymer backbones. A polar nanofibre, owing to the unevenly distributed charge on its

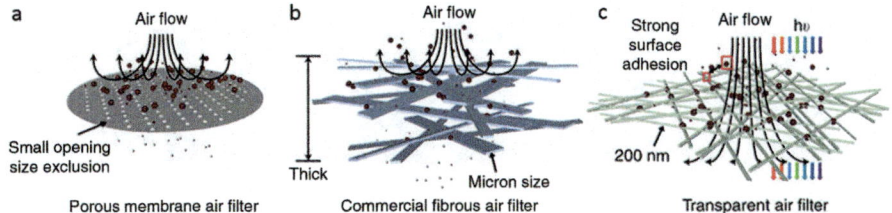

Fig. 9.4 Comparison of different air filters. (**a**) Schematic of a porous air filter capturing PM particles by size exclusion. (**b**) Schematic of a bulky fibrous air filter capturing PM particles by constraint and adhesion. (**c**) Schematic of transparent air filter that captures PM particles by strong surface adhesion whilst allowing high light and air penetration

molecular backbone, experiences high interfacial tension when in contact with another polar or non-polar surface of a different phase (when compared to non-polar fibres). This phenomenon makes polar nanofibres particularly suitable for surface and depth filtration [18]. It is possible to calculate the interfacial tension at the boundary of a solid polar nanofibre and a relatively moist particle using the Young's equation for surface free energy (γd), as stated below. The Young's equation describes the balance at the three-phase contact of solid, liquid and gas. However, this equation is not generally used for solid–solid or liquid–liquid interfaces.

$$\gamma_{sv} = \gamma_{sl} + \gamma_{lv}\cos\theta_Y$$

where the γ_{sv}, γ_{sl} and γ_{lv} are the interfacial tensions between the different phases which form the equilibrium contact angle of wetting, often referred to as Young's contact angle, θ_Y.

Polyacrylonitrile (PAN) is a polar polymer that is commonly used in nanofibre air filters. The polar nature and small fibre diameters make PAN nanofibres highly suitable for use in face masks that are required to remove PM2.5 particles. Electrospun nanofibres made from PAN are also strong and can withstand conversion into filters [21]. PM2.5 particles are often electrically charged, thus enabling bonding between the particles and the nanofibres by means of dipole interactions. The PM2.5 particles typically have sticky amorphous carbon-like morphologies with cores containing condensed solid matter and outer surfaces containing light organic matter with polar functional groups (C–O, C=O and C–N). In such situations, filtration materials that have high dipole moments are more efficient at attracting particulate pollutants (Figs. 9.4 and 9.5) [22, 23]. Jing et al. recently showed that high dipole moments can be created on the surface of PAN nanofibres with the addition of an ionic liquid in the form of diethyl ammonium dihydrogen phosphate (DEAP) [23]. The resultant PAN nanofibre was also more hydrophilic with greatly enhanced particle capturing efficiency due to the enhanced surface roughness and the associated increase in the number of adsorption sites. Moreover, the filters also showed high levels of air permeability. Since PAN nanofibres are currently being electrospun for air filtration applications on a commercial scale, this simple modification technique has great potential to be translated into commercial products [22].

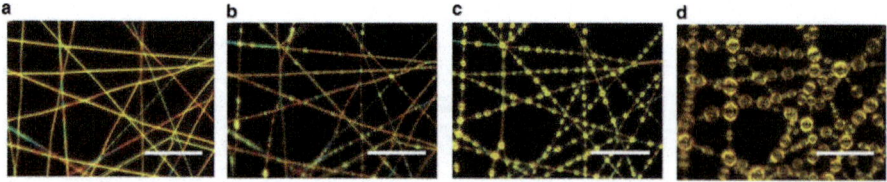

Fig. 9.5 In situ study of PM2.5 particle capture over time using PAN nanofibres. The timescales for (a--b) is 0, 5, 120 and 600 seconds respectively [22]. Reprinted with permissions from reference [22], Copyright 2015, Nature Publishing Group

In general, functional filter membranes work on the principle of creating an affinity for a molecule based on its physical/chemical or biological properties. Relying on the specific ligands immobilised or secondary phases present at the surface, nanofibres capture and separate molecules selectively [24].

In addition to high filtration efficiency, an ideal filter should manifest high breathability. The packing density of the nanofibres greatly influences the breathability of the filtration media. Packing density is often referred to as solidity, which is the volume of solids in the medium per unit volume. Solidity is the opposite value of porosity. The solidity of a nanofibre material (α) is estimated using the following empirical formula:

$$\alpha = W / (\rho Z)$$

where W is the basis weight of the nanofibre filter (mass of the nanofibres per unit filter area), ρ is the density of the fibre material and Z is the thickness of the nanofibre filter mat.

A high nanofibre packing density results in smaller pore sizes and lower breathability, but higher filtration efficiency. Inversely, a low nanofibre packing density results in larger pore sizes and higher breathability, but lower filtration efficiency. It is, therefore, possible to engineer the nanofibre component to best suit the filter performance requirements.

Another important consideration for the design of nanofibre air filters is thermal stability. Some filter applications involve the filtration of hot air or gases (e.g. vehicle exhausts) or even exposure to steam. Zhang et al. recently developed polyimide (PI) nanofibre filters that are able to withstand temperatures of up to 370 °C. These filters were shown to have a 99.98% removal efficiency of 0.3 μm sized particles and achieved the standard requirements of high-efficiency particulate air (HEPA) filters [25]. Wang et al. recently incorporated SiO_2 particles into PI nanofibre membranes, and the resulting filtration media showed high porosity, excellent electrolyte wettability, extraordinary ionic conductivity and outstanding thermal stability [26].

There is also great potential to functionalise the surfaces of nanofibres to create filters with enhanced filtration efficiencies and the selective adsorption of particles. Currently, Kode Biotech, a New Zealand-based bio-surface innovation company, is developing a unique surface coating technology that imparts functionality to

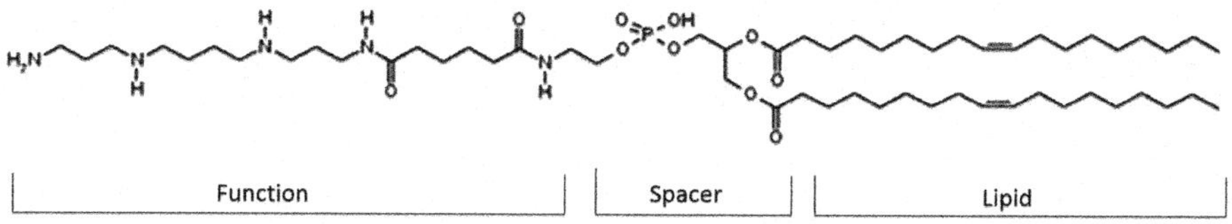

Fig. 9.6 Structure of FSL-Spermine Kode molecule

surfaces in a simple manner without the use of complex solvents [27–29]. Their unique surface coatings consist of synthetic lipid-based molecules called FSLs (Function-Spacer-Lipid), and current work involves the development of FSLs with different functionalities for various applications. Recent collaborative work between Kode Biotech and Revolution Fibres Ltd. has resulted in the development of functional FSL modified nanofibre for air filtration applications. FSL-spermine is one of the molecules that were investigated for its ability to adsorb PM2.5 particles. It is a polycationic molecule that can result in the enhancement of various interactions (mostly dipole interactions) between pollutant particles and the coated surface. The function (F) component of the molecule is responsible for adhering to particles, and the spacer and lipid (S, L) components of the molecule optimise the presentation and activity of the function (F) component as well as enhance the hydrophilicity of the coated surface. The structure of FSL-spermine can be seen in Fig. 9.6.

There are great opportunities for the commercialisation of these new methods of capturing PM 2.5 particles. There is much demand for PM2.5 filtration products in China (especially surgical, PM2.5 and N95 respiratory masks), and the market is substantial and growing. According to Taobao, a popular Chinese online shopping website, China has experienced a dramatic increase in the number of air purifier sales since 2012 (Fig. 9.7). It has also been reported that sales for face masks in China is growing at a rate of 173% per week. The sale of air filtration products is also increasing rapidly in countries such as India, Japan, the USA, South Korea and the UK.

The dominant player in the Chinese market for surgical, N95 and PM2.5 and functional nanofibre respiratory masks is 3M, which maintains a market share of 70–80%. The remaining 20–30% of the market is shared by Makrite, Dasheng, Tianjin Teda, Gangkai, Chaomei, Jinfuyu, Blue Star, Fullstar, Hubei and Xiantao. The market for non-certified face masks is shared by many small and locally owned face mask producers.

It has been noted that the uptake of emerging nanofibre technology has been slow in face mask and air filter applications, and this is likely due to the relatively high cost of the materials as well as a lack of awareness among manufacturers and the consumers. However, some manufacturers have commercialised nanofibre air filter products, including Donaldson Company Inc., Clarcor Inc., Nano and Advanced Materials Institute Ltd. (NAMI), Finetex EnE and Revolution Fibres Ltd.

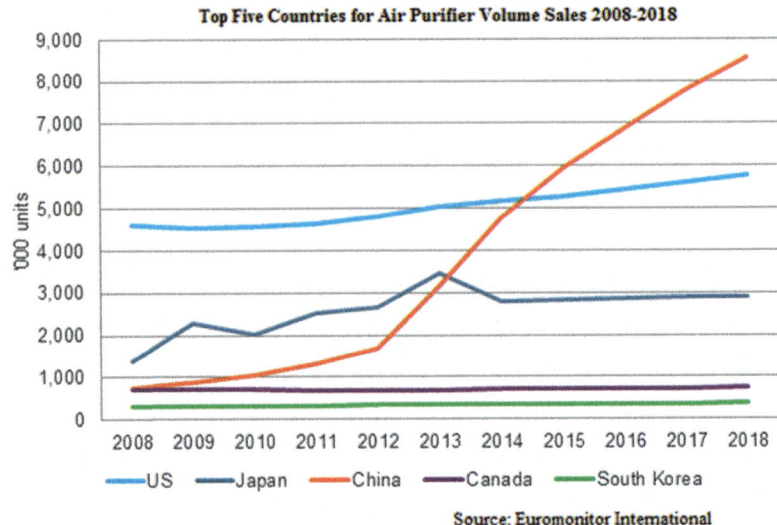

Fig. 9.7 Sales forecasts for consumer air purifiers by Taobao. Source: Euromonitor International, © Financial sense

9.4.2 Volatile Organic Compounds and Toxic Gases

Volatile organic compounds (VOCs) are a wide-ranging group of organic molecules that are considered to be one of the major contributors to air pollution. Sources such as industrial plants, vehicles and aircraft emit these pollutants into the atmosphere causing harm to human health and the ecosystem [30, 31]. International concerns regarding VOCs arise due to their toxic nature, their ability to travel great distances and their tendencies to accumulate and distribute themselves in the environment [32]. Methane is considered to be a less harmful VOC but can contribute to global warming [33] whereas more harmful VOCs such as benzene can lead to photochemical smog when combined with other gases such as nitrogen oxides (NOx). A photochemical smog is the chemical reaction of sunlight, NOx and volatile organic compounds (VOCs) in the atmosphere, which results in the formation of airborne particles (particulate matter) and ground-level ozone [34].

Activated carbon has traditionally been the material of choice for capturing VOCs and photochemical smog. Most of the commercially available VOC filtration materials contain activated carbon. In recent times, cyclodextrins have been utilised by researchers to modify electrospun nanofibre materials to enhance the capture of VOCs. Cyclodextrins are produced from starch, are naturally occurring and are non-toxic cyclic oligosaccharides. Cyclodextrins are a family consisting of three major compounds: α-cyclodextrin (α-CD), β-cyclodextrin (β-CD) and γ-cyclodextrin (γ-CD). Each of the cyclodextrins is crystalline, homogeneous and non-hygroscopic substances which consist of torus-like macro-rings built up from glucopyranose units: α-CD is comprised of six glucopyranose units, β-CD is comprised of seven such units and γ-CD is comprised of eight such units. Because of their complex structures, cyclodextrins are able to become hosts for intermolecular interactions and therefore are able to form non-covalent inclusion complexes with various other molecules. They are recognised for their complexation capacities with hazardous chemicals and organic molecules and are used in many industrial applications [35, 36]. Kayaci et al. have incorporated all three major forms of cyclodextrins into polyester (PET) nanofibres for entrapment of aniline vapor [37]. They showed that electrospinning PET with cyclodextrins also improved the mechanical properties and produced bead-free nanofibres. γ-CD-functionalized PET nanofibres exhibited the greatest aniline capture efficiency when compared to those functionalized with α-CD and β-CD. Previously, researchers had successfully functionalized poly (methyl methacrylate) PMMA nanofibres with different concentrations of β-CD to entrap aniline, styrene and toluene volatiles [38].

Celebioglu et al. was the first to report the electrospinning of cyclodextrin inclusion complexes (CD-IC) without using a carrier polymer matrix and were successful in electrospinning hydroxypropyl-β-cyclodextrin (HPβCD) and its inclusion complexes with triclosan (HPβCD/triclosan-IC) [39]. Further tests were performed to compare the molecular entrapment of volatile organic compounds (aniline and benzene) by HPβCD and HPγCD [31]. It was found that HPβCD was more successful at encapsulating the VOCs than was HPγCD, and both compounds showed a higher affinity for the encapsulation of aniline over that of benzene. A summary of the cyclodextrin functionalization research for VOC capture can be seen in Table 9.1.

Another emerging and commercially scalable technology for the removal of VOCs involves the use of Metal Organic Framework particles or MOFs. MOFs have incredibly high surface areas and their properties can be modified to enable them to capture different VOCs such as xylene, benzene, chloroform and styrene, toxic gases such as carbon monoxide (CO), oxides of nitrogen (NO_x) and oxides of sulphur (SO_x) and chemical warfare agents such as Soman (GD), mustard gas (HD), Sarin (GB) and Lewisite (L).

While traditionally used activated carbons have a proven track record in the capture of VOCs, one shortcoming is the difficulty in attaching functional molecules to the carbon to enable board spectrum protection. Since MOFs allow for the covalent anchoring of functional groups, they can be extensively used in the capture of various classes of chemicals. MOF technology is gaining commercial traction, and NuMat Technologies and the Linde group have formed a collaborative partnership to develop next-generation separation and storage technologies based on MOFs for the industrial-gas sector.

The combination of MOF and nanofibre technologies could result in a new generation of functionalised air filter materials that can remove particulate matter as well as VOCs.

It has been shown that the surface area of nanofibres can be doubled by incorporating MOFs into the material, thus enhancing the filtration efficiency and active surface area of the nanofibres. Zhang et al. analysed and tested the performance of PAN nanofibres with ZIF-8 MOFs. The BET surface area of the PAN nanofibre was improved from 115 to 1024 m^2/g with the incorporation of ZIF-8 nanoparticles at a

Table 9.1 Summary of the cyclodextrin functionalization research for VOC capture

Sample	Solvent	%CD	Average fibre diameter (nm)	Target molecule	Reference
HPβCD/ water-NF	Water	160	745 ± 370	Aniline and benzene	Celebioglu and Uyar [39]
HPβCD/ DMF-NF	DMF	120	1125 ± 360	Aniline and benzene	
HPγCD/ water-NF	Water	160	1165 ± 455	Aniline and benzene	
HPγCD/ DMF-NF	DMF	125	2740 ± 725	Aniline and benzene	
PMMA/β-CD	DMF	10	675 ± 89	Aniline styrene and toluene	Uyar et al. [38, 40]
PMMA/β-CD	DMF	25	625 ± 70	Aniline styrene and toluene	
PMMA/β-CD	DMF	50	816 ± 77	Aniline styrene and toluene	
PET/α-CD	TFA/ DCM	25	900 ± 560	Aniline	Kayaci and Uyar [37]
PET/β-CD	TFA/ DCM	25	830 ± 510	Aniline	
PET/γ-CD	TFA/ DCM	25	790 ± 490	Aniline	

mass ratio of 60% [41]. The authors predicted that the defects and unbalanced metal ions on the surfaces of the MOFs offer positive charges that can polarise airborne particles. This results in increases in electrostatic interactions which can enhance the capture of SO_2 and particulate matter. When tested in hazy environments in Beijing (PM2.5 = 350 $\mu g/m^3$, PM10 = 720 $\mu g/m^3$, RH = 58.6% and T = 23.5 °C), the ZIF-8/PAN MOF/nanofibre composite outperformed nanofibre filters with other MOFs or no MOFs in terms of PM2.5 and PM10 particle removal efficiency. The ZIF-8/PAN composite filter layer was shown to selectively adsorb toxic gases such as SO_2 when exposed to a SO_2/N_2 mixture with a pressure drop of 20 Pa and a gas flow rate of 50 ml/min. After SO_2 adsorption, the modified filter was regenerated by exposing it to a stream of N_2 gas, thus showing that MOF-modified filters can also be reused and recycled.

This MOF/nanofibre technology for selective gas or VOC capture could easily be upscaled using electrospinning techniques and could provide the face mask/filtration markets with unique and potentially game-changing protective air pollution control materials.

Other kinds of MOF particles that has been proven to capture VOCs and other toxic gases are MOF-199, Mg-MOF-74 and UiO-66-NH2 [42].

Qian et al. recently reported the synthesis of indium-based MOFs and their ability to adsorb CO_2. Such MOFs are structurally rigid and stable with open metal sites and have great potential to provide active gas adsorption sites in electrospun nanofibre materials [42]. Furthermore, Vellingiri et al. analysed three MOFs (MOF-5, Eu-MOF and MOF-199) to determine the adsorption capacities of a mixture of 14 volatile and semi-volatile organic gaseous species. The results showed that MOF-

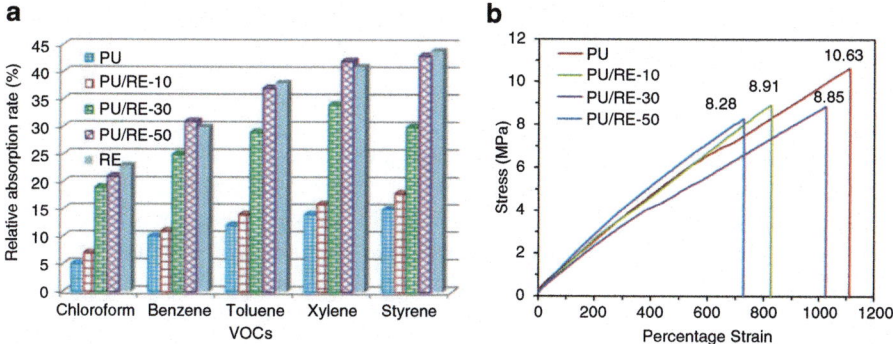

Fig. 9.8 (**a**) VOC adsorption capacity of differently modified PU nanofibres; (**b**) Tensile strength of PU nanofibres with different amounts of RE [30]. Reprinted with permissions from reference 30, Copyright 2017, MDPI open access

199 had the highest equilibrium adsorption capacity of the three MOFs and showed the presence of strong π–π interactions. It was also noted that the polarity of the guest molecule greatly affected the absorption behaviour of the MOFs [43].

Rare earth metal oxides (RE) are also evolving as active materials that are suitable for the removal of VOCs. Some of these oxides include Ce_2O_3, CeO_2, La_2O_3, Pr_2O_3, Nd_2O_3, Sm_2O_3 and Er_2O_3. Ge and Choi recently utilised rare earth oxide powders to modify nanofibres [30]. The authors showed that the addition of RE powders increased the tensile strength of the nanofibre mats, and more importantly, increased the adsorption efficiency of several different VOCs. As can be seen in Fig. 9.8, increases in the concentration of RE in the nanofibre resulted in greater volatile absorption rates, but the authors also indicated that higher RE concentrations caused nozzle blockages during the electrospinning process. This successful functionalisation of nanofibres with the addition of RE suggests great potential for this approach and the need for further investigation in this field.

Bioremediation is another interesting technology that can be utilised for VOC removal. Bioremediation involves the biodegradation of organic contaminants into carbon dioxide, water, inorganic compounds and cell proteins, and also includes the transformation of complex organic contaminants to other simpler organic compounds by microbial activity [44]. Blasi et al. in 2016 tested 163 fungal strains, mostly members of the dematiaceous fungi group, to determine whether or not the fungi could utilise VOCs such as toluene, hexadecane and PCB 126 as the sole carbon and energy source for growth. The results were promising, but this research is still in its infancy and has yet to be fully explored [45]. Bacterial species such as Pseudomonas spp., Sphingomonas spp., Flavobacterium spp., Burkholderia spp. (Gram-negative), Rhodococcus spp., Mycobacterium spp. and Bacillus spp. (Gram-positive) have successfully been used to degrade polycyclic aromatic hydrocarbons (PAHs) and VOCs, including naphthalene, phenanthrene, pyrene, anthracene and benzopyrene [46]. However, bioremediation processes have yet to be combined with nanofibres for active filtration applications, and there appears to be great scope for further work in this field.

9.5   Recent Developments in Nanofibre Filter Media for Water Filtration

Contaminants in water can come from many different sources such as industry, waste treatment plants, pesticides and rainwater run-off. The pollutants can be organic, inorganic, radioactive or acidic/basic in nature [47] which can have severe health impacts on humans such as diarrhoea, tuberculosis, dysentery, cholera, etc. The World Health Organization has reported that 842,000 deaths occur every year due to diarrhoea in 2012 [48]. Water pollution can be considered a major factor in the realisation of these grim statistics. Since the 1990s, water pollution has worsened in almost all rivers in Latin America, Africa and Asia. Of the 1.1 billion people without access to improved water sources worldwide, around 84% live in rural areas [49]. In such areas where the provision of normal water supply infrastructure is lacking, water treatment technologies that are costly and energy intensive are impractical and are not suitable. In such scenarios, nanotechnology can be used in the development of affordable, low-energy and low-maintenance water filtration systems. Electrospun nanofibre filtration media is being increasingly used and modified in different ways to suit the demands of water filtration applications for the removal of dyes, heavy metals, particles, organic compounds and microbes [50].

The global market for water purifiers was worth US 53.35 bn in 2016 and is estimated to double by 2025 according to recent market research [51].

The water filtration industry already has several dominant players in the market offering nanofibre membrane-containing products. These are mostly for the filtration of water from non-potable sources such as sea water or standing water and include DuPont's Hybrid Membrane Technology, Nitto Denko's nanomembranes for water desalination, Hollingsworth and Vose's, NanoWeb for liquid filtration, HiFyber's liquid filtration membranes and Finetex's Technoweb™ Filtrepro Series of liquid filters. However, liquid filters containing nanofibre membranes designed for the capture of specific pollutants such as heavy metals, dyes, pathogens and radioactive contaminants have yet to be commercialised. Further research and development of nanofibre materials that can capture and remove industrial pollutants and heavy metals from polluted river water (some listed below) would greatly benefit governments and save millions of lives from diseases. Several new approaches for the removal of contaminants from drinking water are listed below [44].

9.5.1   Dye Removal

It is estimated that 5000 tons of dye are discharged into the environment annually as a result of increasing industrialization. Industries involved in the manufacture of dyestuff, paper, leather, cosmetics and textiles are generally responsible for the majority of discharge of harmful organic and inorganic dyes into fresh water systems. These pollutants typically have high turbidity, contain suspended toxic compounds and oxidants and pose significant threats to aquatic and human life if untreated [52, 53]. Treatment methods for dye removal from water include adsorption, coagulation and biological treatments [54]. However, dye removal can be challenging due to their complex molecular structures and slow degradation rates [55]. Electrospun nanofibre membranes have the potential to be used in alternative methods of dye removal. Despite having pores that are too large for the size exclusion of dye particles, the large surface areas of the nanofibres and the ability to functionalise the nanofibre surfaces can be utilised for dye removal [56, 57].

Nanofibres synthesised from biopolymers are increasingly gaining favour as adsorbents for dye removal [58–61] due to their environmental and sustainability benefits. Cellulose nanofibres have been successfully used in the removal of dye particles from water as they contain reactive hydroxyl (–OH) functional groups on their surfaces that adhere to dye molecules [62]. A commonly used method for cellulose membrane modification involves an oxidative surface treatment using (2,2,6,6-tetramethyl-piperidin-1-yl)oxyl (TEMPO) [62–64]. However, TEMPO is a toxic substance and thus has limited suitability for large-scale use. Recently, a more environmentally friendly solvent-free method has been reported using Meldrum's acid to modify cellulose-based polyvinylidene fluoride (PVDF) membranes. This treatment resulted in the enhanced adsorption of crystal violet, a positively charged (cationic) dye, from contaminated water [65]. Furthermore, Aziz et al. reported the removal of negatively charged (anionic) dyes, using silk fibroin (SF)/polyacrylonitrile (PAN) double-layer nanofilters with the addition of polyaniline (PANI)/TiO_2 nanoparticles as modifying agents [66]. The nanoparticles were also found to increase the strength and elasticity of the fibres in the nanofilters.

Amine-functionalization is also an effective modification treatment for nanofibres due to the pollutant-chelating capacity of the amine functional groups [67]. For example, cobalt-ferrite nanofibres (Co-ferrite NF) were recently modified by L-arginine to complex with both anionic and cationic dyes. It was shown that the adsorption of the anionic dye decreased as the pH was increased due to reductions in electrostatic attractive forces. Alkaline conditions also encouraged the adsorption of a cationic dye due to the negative charge of the adsorbent surface. Cyclodextrins have also been utilised for dye removal [36, 68], and Chen et al. in 2011 applied a simple filtration method for the functionalization of carbonaceous nanofibres with β-CD for the removal of fuchsine acid (Fig. 9.9) [36].

9.5.2   Heavy Metal Removal

Heavy metals are naturally occurring elements that have a high atomic weight and density. They are important constituents of several key enzymes and play important roles in various oxidation–reduction reactions. However, they carry major health risks, if found in excess. Some toxic heavy metals include arsenic, lead, mercury, cadmium, chromium, aluminium and iron [69, 70].

Current methods for purifying heavy metals ions are based on ion-capturing mechanisms. These methods are typically expensive, and there is a great need for

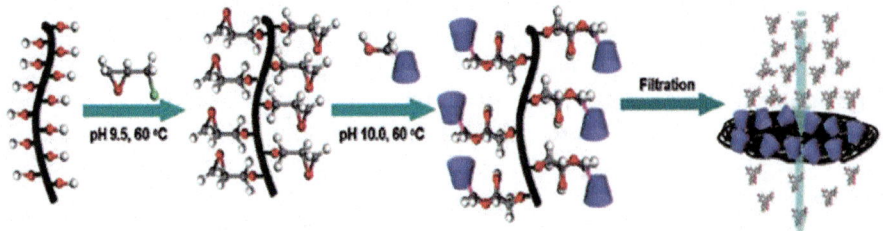

Fig. 9.9 Illustration of the functionalization of a carbonaceous nanofibre membrane by β-CD [36]. Reprinted with permissions from reference 35, Copyright 2011, American Chemical Society

new cost-effective approaches. ETH Zurich researchers have developed a low-cost amyloid-carbon hybrid membrane from denatured whey protein and activated carbon for filtering out heavy metal ions and nuclear waste from water [71]. This technology has great potential to be commercialised. However, most of the commercial water filters in existence are lacking in filtration efficiency that can be improved with the addition of smaller-diameter fibres such as electrospun nanofibres.

Molecular imprinted polymers (MIPS) are polymer particles that are created to have a "memory" of the shape and the functional groups of a template molecule. They can be designed to have a high selectivity and affinity for a target molecule and can therefore be used in liquid or air filtration applications to attract and bind to contaminants. MIPs are synthesised by complex formations between template molecules and functional monomers by either covalent or non-covalent interactions [72] and provide the advantage of high selectivity and specificity towards a given analyte [73]. However, the presence of significant hydrophobic interactions between the MIPs and the template molecules leads to hydrophobically driven nonspecific bindings in pure aqueous media. To solve this problem, G. Pan et al. synthesised MIPS with ultrathin hydrophilic shells that showed greater water compatibility [74]. Luo et al. in 2017 reported the preparation of Pb(II) ion-imprinted polymers with hydrophilic bi-component polymer brushes. The polymer showed high Pb(II) specificity as well as anti-clogging properties [44]. Arsenic and selenium removal have also been reported with the use of MIPs [75]. Not many studies have been performed to date to investigate the functionalization of electrospun nanofibres with MIPs, especially for the removal of heavy metals from waste water. This is seen as an emerging technology with great potential, and further research in this area is anticipated. Ligar Polymers Ltd is one of the first companies in the world to explore large-scale commercial applications for MIPs. It is currently developing MIPs with an affinity for strontium, with the aim of producing a viable method for removing radioactive strontium from Fukushima harbour in the wake of the 2011 Fukushima nuclear disaster. In addition to this, Ligar has also synthesised chromium selective MIPs for wastewater treatment.

Charge modification is a commonly used technique to prevent heavy metal particles from binding to membranes. For example, Mohamed et al. in 2017 used amines to modify TiO_2 nanoparticles to carry a positive charge [76]. They further

Fig. 9.10 SEM images of (**a**) PAN nanofibres and (**b**) PAN-CNT/TiO$_2$-NH$_2$ [76]. Reprinted with permissions from reference [35], Copyright 2017, Elsevier

used these modified nanoparticles to functionalize PAN/CNT for chromium removal from wastewater (Fig. 9.10). Chromium exists in two oxidation states, namely Cr(VI) and Cr(III). Cr(VI) is more toxic, carcinogenic and mutagenic to living organisms when compared to Cr(III) [77, 78]. The charge-modified PAN/CNT showed a remarkable removal proficiency for Cr(VI). Moreover, the removal was predicted to be based on both mechanisms of adsorption and reduction, which is noteworthy as the reduction of Cr(VI) to Cr(III) can decrease the toxicity to a great extent. In addition to this, Mohamed et al. also showed that the adsorption capacity remained at 80% even after five times usage, indicating a good reusability factor. The researchers had further success in using the same membrane for arsenic As(III) and As(V) removal [79]. However, SEM imaging showed that the pores of the membrane were largely blocked after modification, indicating that despite the good heavy metal affinity, they would have poor permeability if used as filtration membranes.

Furthermore, Kumar et al. [80] recently applied a simple oxidation modification technique to polyaniline/g-C$_3$N$_4$ nanofibres using a ternary mixture of H$_2$SO$_4$, HNO$_3$ and H$_2$O$_2$. The nanofibres developed a net positive charge on their surfaces which provided the resultant nanofibre composite with selective binding capacity for Cr(VI). Oxidation with TEMPO can also enhance the absorption affinity of Copper (II), as the introduced carboxylate groups have an affinity for Cu++ ions. Kumar et al. further demonstrated that these oxidised membranes could also be used for other metals such as Ni (II), Cr(III) and Zn(II) [80]. However, as discussed previously, the toxicity of TEMPO should be taken into consideration.

A summary of recently published methods for the removal of dyes and heavy metals from water can be seen in Table 9.2.

9.5.3 Antimicrobial Removal and Antifouling Effects

Microbial pathogens are responsible for a large number of diseases in developing and developed countries in the world. According to the World Health Organization, a large percentage of children under the age of five living in Asian and African

Table 9.2 Summary of recently published methods for the removal of dyes and heavy metals from water

Surface	Modifying agent	Target	Reference
PAN/silk fibroin	Polyaniline/TiO$_2$	HFGR dye	[66]
PVA	DETA (diethylenetriamine) EDA (ethylenediamine)	Direct Red 23 (DR23) Direct Blue (DB78)	[60]
PAN	Diethylamine, diethylenetriamine, triethylenetetramine	Direct Red 23 (DR23) Direct red (DR80)	[81]
Co-ferrite NF	L-arginine	DR80 and BR46	[82]
Cellulose based PVDF	Meldrum's acid and carboxyl groups	Crystal violet	[65]
PAN/CNT	TiO$_2$-NH$_2$	Arsenic	[79]
PAN/CNT	TiO$_2$-NH$_2$	Cr(VI)	[76]
Polyaniline/g-C$_3$N$_4$	H$_2$SO$_4$-HNO$_3$-H$_2$O$_2$	Cr(VI)	[80]
PAN	Amidino diethylenediamine	Cu(II), Ag(I), Fe(II) and Pb(II)	[83]
Cellulose NF	PAA (polyacrylic acid), PIA (poly cationic acid)	Cd(II)	[84]
Cellulose and chitin NF	TEMPO	Cu(II), Ni(II), Cr(III) and Zn(II)	[85]

countries are affected by microbial diseases transmitted through water [86]. In the USA, it is estimated that 560,000 people suffer from severe waterborne diseases and 7.1 million people suffer from mild-to-moderate infections each year resulting in an estimated 12,000 deaths per annum [87]. The United States Environmental Protection Agency has identified over 500 waterborne pathogens that pose potential health risks to humans [88]. Different pre-functionalized membranes are used for the detection and removal of pathogens, where the modification agents are either simply immobilised or covalently bonded to the membranes [89]. Varieties of nanofibres have been electrospun and studied for the purpose of pathogen removal. Some active agents that have been used to modify nanofibre surfaces include metal nanoparticles, antibiotics, triclosan, chlorhexidine and biguanides [90].

The incorporation of silver nanoparticles (AgNP) into filtration membranes can result in good antibacterial properties. The use of AgNPs has been gaining a lot of recent attention as they are very effective antimicrobial agents, even when used at lower concentrations. AgNPs are also considered to be non-toxic, have FDA approval, are widely accepted in filtration products and are easy and inexpensive to synthesise [90]. PLGA-chitosan mats that were modified with graphene-oxide-silver nanocomposites were found to have antibacterial properties against three types of microorganisms [91], namely Escherichia coli, Pseudomonas aeruginosa and Staphylococcus aureus. The blending of chitosan with PLGA resulted in a more homogenous polymer surface but, more importantly, provided amine functional surface groups that were able to bond with the carboxyl groups of graphene oxide. Moreover, the PLGA and chitosan blend produces nanofibres with good mechanical properties that are also biodegradable in nature. Dubey et al. in 2015 [92] also encapsulated AgNPs into a hydrophilic-hydrophobic polymer blend of poly(ethylene oxide) and polycaprolactone (PCL). Apart from the antibacterial resistance against *E. coli*, the addition of AgNPs also reduced the fibre diameter from an average of 150–300 nm to an average of 70–150 nm and enhanced the surface roughness. A very recent study by Nthunya et al. [93] is worth a mention because of their use of a low-cost Fe powder to supplement the use of the more expensive Ag particles. They used β-cyclodextrin and cellulose nanofibres in the ratio 1:1 polymer concentration to achieve bead-free nanofibres to which they added Fe and Ag particles. The particle-modified nanofibre membrane showed antibacterial activity against 12 pathogenic bacterial strains, showing the greatest antibacterial activity against *E. faecalis* and *P. mirabilis*. As the nanofibres were activated using a greener UV photochemical reduction process in the presence of N2 under ionised water vapor, this can be highlighted as a good example of an environmentally friendly procedure that can be commercialised.

Membrane fouling can cause severe reductions in the flux and quality of filtered water. Cyclodextrins are often used as antifouling agents in membranes as they have the ability to provide hydrophilization [39, 94]. Hydrophilicity is a desired property of a filtration membrane that increases permeation and provides an antifouling effect. However, methods to make membranes more hydrophilic can be complex. Yu et al. in 2015 reported a simple method of dip-coating PVDF membranes in a boron–cadmium solution to make them more hydrophilic [95]. The resultant membranes showed an increase in water flux as well as remarkable antifouling properties. Similar methods can be adopted for functionalizing electrospun nanofibre filtration media.

9.6 Author Insights and Conclusions

Increases in human population, industrial growth and environmental destruction have caused pollution in the air and water throughout the world. In recent years, humanity has become increasingly aware of the growing health risks associated with this pollution which can affect life at work, in the home and throughout our daily lives. The role of filtration is an important preventative measure to help purify and decontaminate two of the vital necessities of life, namely air and water. It is an incredibly important industry affecting almost all facets of life and industry. Nanofibre air filters have been on the market for a number of years and date back as far as the 1930s.

It is a fascinating story of a "forgotten" technology riding the wave of "nanotechnology" and becoming rediscovered in research circles. Now, the nanofibre is one of the fastest growing and the most widely adopted form of nanotechnology. While, therefore, the concept is nothing new, the filtration industry is noticing an upsurge in growth of nanofibres as the demands for clean water and air intensify. As the awareness of health outcomes from pollution increases, so too do the demands for greater protection. Nanofibres are now being widely accepted as being particularly effective in stopping submicron pollutants with a minimal impact on pressure drop and energy.

The nanofibre industry is not without its challenges. Electrospinning is probably the most successful technique to fabricate nanofibres both in lab and mass scale. To be commercially viable, the electrospinning process must be able to produce nanofibres at large enough volumes and for a low enough cost to be commercially viable when competing with existing alternative technologies or products. In order to compete with existing textiles (and specialty films), nanofibres must offer distinct performance or production advantages such as

Improvements in surface area, porosity and reduced thickness and/or weight
Use of materials and functional additives that would not withstand traditional fibre-
    making techniques (e.g. melt blowing or extrusion)
Textiles or membranes small enough to be used in miniaturised devices (e.g.
    personal filters, electronics, medical devices, etc.)

Nanofibre air filters have been on the market for a number of years. Main producers include Donaldson's Ultraweb®, Exceed, eSpin Technologies and Hollingsworth & Voses' Nanoweb® filter media and the markets typically devoted to HVAC and industrial requirement. Within the "PM 2.5 facemasks" markets, the landscape for nanofibre media suppliers is getting increasingly more competitive. Standard polymeric nanofibres are usually focused on achieving a 99% filtration efficiency for particle sizes ranging from 1 microns to 10 microns, but additional plies (layers) are required to sufficiently remove smaller, lung-damaging particles. However, the addition of extra layers typically results in reduced breathability and comfort. Nanofibres can offer greater protection with less airflow resistance. Nanofibre masks include the Respilon mask, NASK from NAMI and Finetex's Technoweb masks. Incumbent products such as VOG Mask and Metamasks are also introducing nanofibres into their range. Liquid filters using nanofibres are also not new with Dupont and Esfil Tehno (Estonia) offering a diverse range of filters and membranes. Recently, a start-up company Liquidity Nanotech raised significant capital through crowdfunding, offering the "Naked" filter to the market, which is essentially a drinking straw that filters contaminants from water using a nanofibre membrane.

A list of polymers used in filtration is given below in Table 9.3.

In the author's opinion, the next phase of growth is in functional, "intelligent" filters that selectively capture pollutants of concern (VOCs, microbes and viruses, heavy metals, etc.) using active ingredients. Examples of functional additives include nanoparticles (silver, platinum), metal-organic frameworks, molecular imprinted polymers and bioactives. These actives are highly effective but are not widely used in filtration as they do not yet exist in a textile (non-woven) form. This makes them difficult to convert into filters and exploit the high surface areas that they naturally possess. They also solve the growing problem of overloading and biofouling of filters by extending the life of the filter and improving energy efficiency.

Wherever filtration is required, the nanofibre has a significant role to play. Using finer fibres to create greater breathability or use less energy is an appealing factor.

Table 9.3 List of common polymers commercially used in air and water filtration

Polymer	Filtration uses	Properties	Temperature maximum	Author's comments
PA66 Nylon 6,6	High-temperature air and liquid filtration	Extremely cohesive nanofibre	264 °C	PA66 is a commonly used fibre in commercial filtration applications as it is readily scalable, insoluble in water and achieves a high filtration efficiency with relatively low depositions. These properties of PA66 make it useful for both air and liquid filtration at high temperatures. The main limitation of PA66 is the pressure drop being relatively high when compared with polymers such as PMMA due to its cohesiveness [96]; however, this also lends durability to the resulting nanofibre filter
PMMA Poly(methyl methacrylate)	High airflow/ low pressure drop Air and liquid filtration [97]	Extremely lofty nanofibre	160 °C	PMMA is readily scalable, plus due to its insolubility in water, it is useful in both air and water filtration. PMMA is incredibly lofty, which allows for high filtration efficiency with very little pressure drop when compared with PA66 or more cohesive polymers. It is suitable for face masks or low-pressure applications which require maximum breathability. The main limitation is its relatively low durability compared with more cohesive nanofibres such as PA66
PVOH Polyvinyl alcohol	Air filtration and liquid filtration (when cross-linked) [96]	Cohesive nanofibre	200 °C	PVOH being soluble in water proposes a challenge in filtration/scale-up, as suitability for filtration is limited to dry/non-humid air. However, PVOH can be readily cross-linked (by maleic acid and vitriolic acid as catalyst) to be used in both air and liquid filtration [96]

(continued)

Table 9.3 (continued)

Polymer	Filtration uses	Properties	Temperature maximum	Author's comments
PCL Polycaprolactone	Low temperature filtration [98]	Cohesive nanofibre	60 °C	The main benefit being its biodegradability and insolubility, PCL can be used for both air and water filtration. However, PCL cannot be used in high temperature conditions due to the low melting point
PEO Polyethylene oxide	Low temperature air filtration [99]	Slightly lofty nanofibre	65 °C	PEO having a low melting point and water solubility means usage/scalability is limited; it can however be utilised in air filtration. PEO is more commonly used in combination with other polymers to create a composite nanofibre mat
PVDF Polyvinylidene fluoride	Air and liquid filtration. Clothing filtration [50]	Extremely cohesive nanofibre	177 °C	PVDF is scalable and already utilised in many water filters due to its insolubility, inertness and porosity. PVDF can be used with corrosive solutions and in relatively high temperatures, making it extremely useful for filtration. Furthermore, PVDF readily diffuses water vapour without a loss in filtration efficiency
TPU Thermoplastic polyurethane	Air and liquid filtration [50, 100]	Highly elastic nanofibre	Ranges from 55 to 220 °C based on grade/type	TPU is currently used in some commercial face masks; however, due to use of toxic solvents for electrospinning it is hard to produce the polymer on an industrial scale. Generally, the melt-blown electrospinning technique is used for TPU nanofibres TPU has a moderate pressure drop and is extremely elastic. The resulting nanofibre mat is durable due to its adhesion and elasticity. The limitations of TPU are recognised in high pressure applications due to air/liquid overcoming elastic forces which opens and increases pore sizes

(continued)

Table 9.3 (continued)

Polymer	Filtration uses	Properties	Temperature maximum	Author's comments
PVAc Polyvinyl acetate	Low temperature air filtration [101]	Lofty nanofibre	60 °C	PVac can be effective for air filtration [101]; however, high temperatures must be avoided due to the low melting point. Due to its limitations, it is not widely used
PES Polyether sulfone	Air and liquid filtration [102]	Slightly lofty nanofibre	220 °C	PES is effective for high temperature filtration due to high melting point. PES is highly resistant to most of the solvents and also achieves a high filtration efficiency and low pressure drop
PAN Polyacrylonitrile	High temperature air and liquid filtration [97, 103]	Cohesive nanofibre	>300 °C	PAN is already utilised within industry and is scalable. It is highly porous and used within membrane filters. It is also stable at high temperatures, making it desirable for many applications. It is scalable and used commonly for membrane filters for biomedical purification of fluids
Cellulose acetate	Air and liquid filtration [104]	Slightly lofty nanofibre	170–240 °C	Cellulose acetate is challenging to scale, due to the wide range of different molecular weights and grades that change the resulting nanofibre mat. It can be suitable for high flux water filtration [104], however unsuitable for applications involving concentrated acids or alkalis

As scale increases and costs are reduced, nanofibres will become increasingly used. Electrospinning will face challenges from other processes such as extrusion and melt-blowing, where advances are being made to produce smaller-diameter fibres. Whatever the method, the nanofibre is an outstanding technology for filtration with a lot of scope for future development—all for the benefit and safety of mankind.

Acknowledgements The authors would like to thank Prof. Steve Henry, CEO of Kode Biotech at the Auckland University of Technology for his input and suggestions. We would also like to thank Callaghan Innovation, New Zealand for its extended support and funding.

References

1. Dai J et al (2015) Ambient air pollution, temperature and out-of-hospital coronary deaths in Shanghai, China. Environ Pollut 203:116–121
2. Qiu H et al (2015) Air pollution and mortality: effect modification by personal characteristics and specific cause of death in a case-only study. Environ Pollut 199:192–197
3. Kampa M, Castanas E (2008) Human health effects of air pollution. Environ Pollut 151:362–367
4. Barhate R, Ramakrishna S (2007) Nanofibrous filtering media: filtration problems and solutions from tiny materials. J Membr Sci 296:1–8
5. Pulliero A et al (2015) Genetic and epigenetic effects of environmental mutagens and carcinogens. Biomed Res Int 2015
6. Kannan B, Cha H, Hosie IC (2016) Electrospinning—commercial applications, challenges and opportunities. In: Fakirov S (ed) Nano-size polymers: preparation, properties, applications. Springer International, Cham, pp 309–342
7. Doshi J, Reneker DH (1995) Electrospinning process and applications of electrospun fibers. J Electrost 35:151–160
8. Bhattarai P et al (2014) Electrospinning: how to produce nanofibers using most inexpensive technique? An insight into the real challenges of electrospinning such nanofibers and its application areas. Int J Biomed Adv Res Online J 5:2229–3809
9. Hayes TR, Hosie IC (2015) Turning Nanofibres into products: electrospinning from a manufacturer's perspective. In: Electrospinning for high performance sensors. Springer, Cham, pp 305–329. https://doi.org/10.1007/978-3-319
10. Teo WE, Ramakrishna S (2006) A review on electrospinning design and nanofibre assemblies. Nanotechnology 17:R89–R106
11. Margareth G (2013) Global market and technologies for nanofibres. BCC Research
12. Chen Z et al (2013) China tackles the health effects of air pollution. Lancet 382(9909):1959–1960
13. Christine O (2013) Unearthed, in Unearthed: Greenpeace http://energydesk.greenpeace.org/2013/12/12/map-health-impact-chinas-coal-plants/. Greanpeace, UK
14. Quigley JT (2013) Chinese government will spend $277 billion to combat air pollution. In: The Dipolmat. The Dipolmat
15. Andrew M (2014) U.S. Indoor Air Quality Market, B. Reaserch, Editor. United Kingdom
16. Ravindra K, Sokhi R, Van Grieken R (2008) Atmospheric polycyclic aromatic hydrocarbons: Source attribution, emission factors and regulation. Atmos Environ 42:2895–2921
17. Wania F, Mackay D (1996) Tracking the distribution of persistent organic pollutants. Environ Sci Technol 30:390A–396A
18. Young Chung H (2007) Donalson Co., I., Book review: "Electrospinning of micro and nanofibers: fundamentals in separation and filtration processes", in 2008 special bulletin. Journal of engineered fibres and fabrics: Minneapolis, p 2
19. Mukhopadhyay A (2010) Pulse-jet filtration: an effective way to control industrial pollution Part II: process characterization and evaluation of filter media. Text Prog 42:1–97
20. Spurný KT (1998) Advances in aerosol filtration, p 533
21. Yao J, Bastiaansen C, Peijs T (2014) High strength and high modulus electrospun nanofibers. Fibers 2:158–186
22. Liu C et al (2015) Transparent air filter for high-efficiency PM2.5 capture. Nat Commun 6:6205
23. Jing L et al (2016) Electrospun polyacrylonitrile–ionic liquid nanofibers for superior PM 2.5 capture capacity. ACS Appl Mater Interfaces 8:7030–7036
24. Homaeigohar S, Zillohu AU, Abdelaziz R, Hedayati MK, Elbahri M (2016) A novel nanohybrid nanofibrous adsorbent for water purification from dye pollutants, p 16
25. Li L et al (2017) Three-layer composite filter media containing electrospun polyimide nanofibers for the removal of fine particles. Fibers Polym 18:749–757

26. Wang Y et al (2017) A nano-silica modified polyimide nanofiber separator with enhanced thermal and wetting properties for high safety lithium-ion batteries. J Membr Sci 537:248–254
27. Barr K et al (2016) Biofunctionalizing nanofibers with carbohydrate blood group antigens. Biopolymers 105(11):787–794
28. Blake D, Bovin N, Bess D, Henry SM (2011) FSL constructs: a simple method for modifying cell/virion surfaces with a range of biological markers without affecting their viability. J Vis Exp (54):1–9
29. Williams E et al (2016) Ultra-fast glyco-coating of non-biological surfaces. Int J Mol Sci 17(1):E118
30. Ge J, Choi N (2017) Fabrication of functional polyurethane/rare earth nanocomposite membranes by electrospinning and its VOCs absorption capacity from air. Nanomaterials 7:60
31. Celebioglu A et al (2016) Molecular entrapment of volatile organic compounds (VOCs) by electrospun cyclodextrin nanofibers. Chemosphere 144:736–744
32. Srivastava A, Mazumdar D (2011) Monitoring and reporting VOCs in ambient air. In: Mazzeo NA (ed) Air quality monitoring, assessment and management
33. Boucher O et al (2009) The indirect global warming potential and global temperature change potential due to methane oxidation. Environ Res Lett 4:044007
34. Rani B et al (2011) Photochemical smog pollution and its mitigation measures. J Adv Sci Res 2:28–33
35. Szejtli J (1998) Introduction and general overview of cyclodextrin chemistry. Chem Rev 98:1743–1754
36. Chen P et al (2011) Carbonaceous nanofiber membrane functionalized by beta-cyclodextrins for molecular filtration. ACS Nano 5:5928–5935
37. Kayaci F, Uyar T (2014) Electrospun polyester/cyclodextrin nanofibers for entrapment of volatile organic compounds. Polym Eng Sci 54:2970–2978
38. Uyar T et al (2010) Cyclodextrin functionalized poly(methyl methacrylate) (PMMA) electrospun nanofibers for organic vapors waste treatment. J Membr Sci 365:409–417
39. Celebioglu A, Uyar T (2011) Electrospinning of polymer-free nanofibers from cyclodextrin inclusion complexes. Langmuir 27:6218–6226
40. Uyar T et al (2010) Cyclodextrins: comparison of molecular filter performance. ACS Nano 4:5121–5130
41. Zhang Y et al (2016) Preparation of nanofibrous metal–organic framework filters for efficient air pollution control. J Am Chem Soc 138(18):5785–5788
42. Qian J et al (2017) A microporous MOF with open metal sites and Lewis basic sites for selective CO2 capture. Dalton Trans 46(41):14102–14106
43. Vellingiri K et al (2016) Metal organic frameworks as sorption media for volatile and semi-volatile organic compounds at ambient conditions. Sci Rep 6:27813
44. Luo X et al (2017) Selective removal Pb(ii) ions form wastewater using Pb(ii) ion-imprinted polymers with bi-component polymer brushes. RSC Adv 7(42):25811–25820
45. Blasi B et al (2016) Pathogenic yet environmentally friendly? Black fungal candidates for bioremediation of pollutants. Geomicrobiol J 33(3–4):308–317
46. Biswas B et al (2015) Bioremediation of PAHs and VOCs: advances in clay mineral–microbial interaction. Environ Int 85(Supplement C):168–181
47. Modesti M, Boaretti C (2016) Encyclopedia of membranes, p 1–3
48. World Health Organization (2012) Global Health Observatory Data. Retrieved March 22, 2018, from http://www.who.int/mediacentre/factsheets/fs391/en/
49. World Health Organization (2006) Meeting the MDG drinking water and sanitation target : the urban and rural challenge of the decade. Geneva
50. Nasreen SAAN et al (2013) Advancement in electrospun nanofibrous membranes modification and their application in water treatment. Membranes 3:266–284
51. Water Purifier Market (Technology – Gravity Purifiers, RO Purifiers, UV Purifiers, Sediment Filters, and Water Softener; End-User – Industrial, Commercial, and Household; Accessories –

Pitcher Filter, Under Sink Filter, Shower Filter, Faucet Mount, Water Dispenser, Replacement Filter, Countertop Filters, and Whole House Filters) – Global Industry Analysis, Size, Share, Growth, Trends, and Forecast 2017–2025 (2017) In: Technology & Media, USA

52. Doğan M, Özdemir Y, Alkan M (2007) Adsorption kinetics and mechanism of cationic methyl violet and methylene blue dyes onto sepiolite. Dyes Pigments 75:701–713

53. Pan Y et al (2016) Fabrication of highly hydrophobic organic-inorganic hybrid magnetic polysulfone microcapsules: a lab-scale feasibility study for removal of oil and organic dyes from environmental aqueous samples. J Hazard Mater 309:65–76

54. Kandisa RV, Narayana Saibaba KV (2016) Dye removal by adsorption: a review. J Bioremed Biodegr 07:317

55. Gürses A et al (2016) Dyes and pigments. In: Dyes and pigments

56. Gopal R et al (2007) Electrospun nanofibrous polysulfone membranes as pre-filters: particulate removal. J Membr Sci 289:210–219

57. Ma Z, Masaya K, Ramakrishna S (2006) Immobilization of Cibacron blue F3GA on electrospun polysulphone ultra-fine fiber surfaces towards developing an affinity membrane for albumin adsorption. J Membr Sci 282:237–244

58. Hou C et al (2015) Preparation of PAN/PAMAM blend nanofiber mats as efficient adsorbent for dye removal. Fibers Polym 16:1917–1924

59. Qureshi UA et al (2017) Highly efficient and robust electrospun nanofibers for selective removal of acid dye. J Mol Liq 244:478–488

60. Mahmoodi NM, Mokhtari-Shourijeh Z, Ghane-Karade A (2017) Synthesis of the modified nanofiber as a nanoadsorbent and its dye removal ability from water: Isotherm, kinetic and thermodynamic. Water Sci Technol 75:2475–2487

61. Qureshi UA et al (2017) Electrospun zein nanofiber as a green and recyclable adsorbent for the removal of reactive black 5 from the aqueous phase. ACS Sustain Chem Eng 5:4340–4351

62. Stenstad P et al (2008) Chemical surface modifications of microfibrillated cellulose. Cellulose 15:35–45

63. Johnson R (2010) TEMPO-oxidized nanocelluloses: Surface modification and use as additives in cellulosic nanocomposites. Esf Edu

64. Missoum K, Belgacem MN, Bras J (2013) Nanofibrillated cellulose surface modification: A review. Materials 6:1745–1766

65. Gopakumar DA et al (2017) Meldrum's acid modified cellulose nanofiber-based polyvinylidene fluoride microfiltration membrane for dye water treatment and nanoparticle removal. ACS Sustain Chem Eng 5:2026–2033

66. Aziz S et al (2017) Electrospun silk fibroin/PAN double-layer nanofibrous membranes containing polyaniline/TiO2 nanoparticles for anionic dye removal. J Polym Res 24:140

67. Tahaei P et al (2008) Preparation of chelating fibrous polymer by different diamines and study on their physical and chemical properties. Mater Werkst 39:839–844

68. Liu Z-G et al (2017) Efficient removal of organic dyes from water by β-cyclodextrin functionalized graphite carbon nitride composite. ChemistrySelect 2(5):1753–1758

69. Jaishankar M et al (2014) Toxicity, mechanism and health effects of some heavy metals. Interdiscip Toxicol 7:60–72

70. Tchounwou PB et al (2012) Heavy metals toxicity and the environment. EXS 101:1–30

71. Bolisetty S, Mezzenga R (2016) Amyloid–carbon hybrid membranes for universal water purification. Nat Nano 11(4):365–371

72. Shen X, Xu C, Ye L (2013) Molecularly imprinted polymers for clean water: analysis and purification. Ind Eng Chem Res 52(39):13890–13899

73. Vasapollo G et al (2011) Molecularly imprinted polymers: present and future prospective. Int J Mol Sci 12(9):5908–5945

74. Pan G et al (2011) Controlled synthesis of water-compatible molecularly imprinted polymer microspheres with ultrathin hydrophilic polymer shells via surface-initiated reversible addition-fragmentation chain transfer polymerization. Soft Matter 7(18):8428–8439

75. Mafu LD, Mamba BB, Msagati TAM (2016) Synthesis and characterization of ion imprinted polymeric adsorbents for the selective recognition and removal of arsenic and selenium in wastewater samples. J Saudi Chem Soc 20(5):594–605
76. Mohamed A et al (2017) Removal of chromium (VI) from aqueous solutions using surface modified composite nanofibers. J Colloid Interface Sci 505:682–691
77. Guo X et al (2011) High-performance and reproducible polyaniline nanowire/tubes for removal of Cr(VI) in aqueous solution. J Phys Chem C 115:1608–1613
78. Ku Y, Jung IL (2001) Photocatalytic reduction of Cr(VI) in aqueous solutions by UV irradiation with the presence of titanium dioxide. Water Res 35:135–142
79. Mohamed A et al (2017) Surface functionalized composite nanofibers for efficient removal of arsenic from aqueous solutions. Chemosphere 180:108–116
80. Kumar R, Barakat MA, Alseroury FA (2017) Oxidized g-C3N4/polyaniline nanofiber composite for the selective removal of hexavalent chromium. Sci Rep 7:12850
81. Almasian A et al (2016) Surface modification of electrospun PAN nanofibers by amine compounds for adsorption of anionic dyes. Desalin Water Treat 57:10333–10348
82. Almasian A et al (2016) Zwitter ionic modification of cobalt-ferrite nanofiber for the removal of anionic and cationic dyes. J Taiwan Inst Chem Eng 67:306–317
83. Kampalanonwat P, Supaphol P (2010) Preparation and adsorption behavior of aminated electrospun polyacrylonitrile nanofiber mats for heavy metal ion removal. ACS Appl Mater Interfaces 2:3619–3627
84. Chitpong N, Husson SM (2016) Nanofiber ion-exchange membranes for the rapid uptake and recovery of heavy metals from water. Membranes 6:59
85. Sehaqui H et al (2014) Enhancing adsorption of heavy metal ions onto biobased nanofibers from waste pulp residues for application in wastewater treatment. Cellulose 21:2831–2844
86. Vilasrao TD (2017) Bacterial contamination in drinking water: assesing the potabilty of water. Int Edu Appl Sci Res J (IEASRJ) 2
87. Cabral JPS (2010) Water microbiology. Bacterial pathogens and water. Int J Environ Res Public Health 7:3657–3703
88. EPA (2017) Drinking water contaminant candidate list (CCL) and regulatory determination
89. van den Hurk R, Evoy S (2015) A review of membrane-based biosensors for pathogen detection. Sensors (Switzerland) 15:14045–14078
90. Gao Y et al (2014) Electrospun antibacterial nanofibers: production, activity, and in vivo applications. J Appl Polym Sci 131:9041–9053
91. De Faria AF et al (2015) Antimicrobial electrospun biopolymer nanofiber mats functionalized with graphene oxide-silver nanocomposites. ACS Appl Mater Interfaces 7:12751–12759
92. Dubey P et al (2015) Silver-nanoparticle-Incorporated composite nanofibers for potential wound-dressing applications. J Appl Polym Sci 132:1–12
93. Nthunya LN et al (2017) Greener approach to prepare electrospun antibacterial β-cyclodextrin/cellulose acetate nanofibers for removal of bacteria from water. ACS Sustain Chem Eng 5:153–160
94. Cheirsilp B, Rakmai J (2017) Inclusion complex formation of cyclodextrin with its guest and their applications. Biol Eng Med 2:1–6
95. Yu Z et al (2015) Preparation of a novel anti-fouling β-cyclodextrin–PVDF membrane. RSC Adv 5:51364–51370
96. Qin X-H, Wang S-Y (2008) Electrospun nanofibers from crosslinked poly(vinyl alcohol) and its filtration efficiency. J Appl Polym Sci 109(2):951–956
97. Yun KM et al (2010) Morphology optimization of polymer nanofiber for applications in aerosol particle filtration. Sep Purif Technol 75(3):340–345
98. Cooper A et al (2013) Chitosan-based nanofibrous membranes for antibacterial filter applications. Carbohydr Polym 92(1):254–259
99. Han W, Gaofeng Z, DaoHeng S (2007) Electrospun nanofibrous membrane for air filtration. In: 2007 7th IEEE conference on nanotechnology (IEEE NANO)

100. Gibson HS (2007) Patterned electrospray fiber structures. In: Busnaina A (ed) Nanomanufacturing handbook. CRC Press, Boca Raton
101. Matulevicius J et al (2016) The comparative study of aerosol filtration by electrospun polyamide, polyvinyl acetate, polyacrylonitrile and cellulose acetate nanofiber media. J Aerosol Sci 92(Supplement C):27–37
102. Homaeigohar SS, Buhr K, Ebert K (2010) Polyethersulfone electrospun nanofibrous composite membrane for liquid filtration. J Membr Sci 365(1):68–77
103. Saeed K et al (2008) Preparation of amidoxime-modified polyacrylonitrile (PAN-oxime) nanofibers and their applications to metal ions adsorption. J Membr Sci 322(2):400–405
104. Zhou Z (2016) Electrospinning ultrathin continuous cellulose acetate fibers for high-flux water filtration. Colloids Surf A Physicochem Eng Asp 494:21–29

Index

© Springer International Publishing AG, part of Springer Nature 2018
M. L. Focarete et al. (eds.), *Filtering Media by Electrospinning*,
https://doi.org/10.1007/978-3-319-78163-1

Printed by Printforce, the Netherlands